T0320416

Biophysical Economy

This book explores the concept of transforming the current macroeconomic system from one based on continuous growth that doesn't recognize the fundamental importance of Earth's natural support structures, to a system consistent with the basic views of biophysical economics that acknowledges that all real wealth ultimately derives from planetary resources, both renewable and non-renewable. It describes how data and information collected by various institutions, government agencies, and the private sector can be integrated with existing management structures to transform the "continuous growth" economy into an economy that functions within understandable boundaries on a finite planet.

Features

- Stimulates discussions of the feasibility of a biophysical economy.
- Discusses the historical developments of biophysical economics.
- Offers a practical approach to building a biophysical economy.
- Explores the human experience of living in a biophysical economy.
- Emphasizes the fragility of life in the Universe as we know it.

This book is an excellent resource for academics and students studying sustainable development, as well as for professionals working in the private sector and public institutions with an interest in economic planning for a sustainable future.

Applied Ecology and Environmental Management

Series Editor: Steven M. Bartell, Consultant for Stantec, Ecological Modeling Practice and Sven E. Jorgensen, Copenhagen University, Denmark

The ecology is closely tied to environmental science and management. As a result, ecology has developed from a more descriptive to a more quantitative science, to be applicable as support for environmental management decisions. Several ecological sub-disciplines with emphasis on the application aspects have emerged as a result of this development:

- Ecological Modeling
- Ecological Indicators
- Ecotechnology
- Ecological Engineering
- Ecohydrology
- Ecological Economics
- Ecological Informatics

This book series presents the state-of-the-art and an overview of these ecological sub disciplines with emphasis on the application in environmental management. It summarizes all relevant and basic knowledge needed to understand and utilize in environmental science, environmental management, applied ecology and pollution abatement.

Managing Environmental Data
Principles, Techniques, and Best Practices
Gerald A. Burnette

Global Blue Economy
Analysis, Developments, and Challenges
Md. Nazrul Islam and Steven M. Bartell

Memoirs of an Environmental Science Professor
William Mitsch

Ecological Forest Management Handbook, Second Edition
Edited by Guy Larocque

Biophysical Economy
Theory, Challenges, and Sustainability
Steven M. Bartell

For more information on this series, please visit: https://www.routledge.com/Applied-Ecology-and-Environmental-Management/book-series/CRCAPPECOENV?pd=published,forthcoming&pg=2&pp=12&so=pub&view=list

Biophysical Economy

Theory, Challenges, and Sustainability

Steven M. Bartell

CRC Press is an imprint of the
Taylor & Francis Group, an **informa** business

Designed cover image: © iStock Photos, Credit: RapidEye

First edition published 2025
by CRC Press
2385 NW Executive Center Drive, Suite 320, Boca Raton FL 33431

and by CRC Press
4 Park Square, Milton Park, Abingdon, Oxon, OX14 4RN

CRC Press is an imprint of Taylor & Francis Group, LLC

ISBN: 9781032310794 (hbk)
ISBN: 9781032311616 (pbk)
ISBN: 9781003308416 (ebk)

DOI: 10.1201/9781003308416

Typeset in Times New Roman
by Deanta Global Publishing Services, Chennai, India

Contents

Section 3 Synthesis

Preface

A systems ecologist by profession and self-proclaimed biophysical economist by desire perhaps has no right to suggest an initial design for a biophysical economy. But that is precisely the intent of this book. Fortunately, there are no formal requirements for adopting the mantle of a biophysical economist. Years of experience in the analysis and mathematical modeling of complex, dynamic, and non-linear environmental systems seem preparation enough to launch such an undertaking. At least the author is not encumbered by a formal training in traditional mainstream economics – a "science" that has gotten us to the dire real-world circumstances that requires a global change in plans.

This book was stimulated in part by rediscovering the *Limits to Growth* (1972) and the author's associated interests in exploring the possible extinction of humans as the result of planetary economic mismanagement as modeled by Meadows et al. under the auspices of the Club of Rome. About the same time as delving again into the *Limits* book, Graham Turner's quantitative assessment of the realism of the *Limits* projections came to light – where 30 years of data appeared to corroborate the "business as usual" projections of the Meadows et al. (1972) modeling with frightening accuracy. While there is room to quibble with some of the details and assumptions underlying Turner's assessment, the overall analysis remains compelling. Giving Turner the benefit of the doubt and recalling the 2020–2040 timeframe for the predictions of the Meadows version of global collapse, consideration of an alternative and sustainable economic system as laid out in this book seemed timely.

Undertaking the design of a biophysical economy was also stimulated by the seminal work of Geoffrey West regarding universal laws of scale – particularly in relation to corporations and cities. West's explorations into a science of corporations was quite stimulating and guided me into a corresponding technical literature, some of which filtered into this book.

This effort in suggesting a biophysical economy would not have been possible without having discovered the ground-laying work of John Sterman in his treatise on systems thinking and dynamic models for complex business systems. Perusing this tome underscored the similarities, not entirely surprising, in terminology, concepts, and methodologies fundamental to systems analysis and modeling applied to both business and ecosystems.

In outlining this book, I attempted to be comprehensive in identifying topic areas that would have to be addressed in designing and setting up a nation-scale (and eventually global) biophysical economy. The breadth of topics combined with finite time and resources available for researching and writing produced rather shallow expositions evidenced across the chapters, much to my dissatisfaction. For that I apologize. Nonetheless, the resulting volume can be viewed as at least launching the start of the conversation – or exploring whether it is even possible to have a productive conversation, where such conversation might actually matter, on instantiating a biophysical economy.

I confess to an initial naivete in believing that a biophysical economy might readily obtain through palatable alterations to existing institutions and processes in government, finance, and the private sector, accompanied by rational changes in consumerism. Struggling through personal expansion in language and concepts from the domain of systems ecology to include economic systems in writing this book proved my initial belief to be quite superficial and simple-minded. Not to give away the punchline, but the design and implementation of a biophysical (or otherwise sustainable) economy has less to do with my initial beliefs and more to do with the evolutionary limitations of *Homo sapiens* and cultural niche construction when confronted by finitude.

Finally, I wish to thank Charles Hall and Kent Klitgaard for an invitation to review several chapters of the second edition of their treatise, *Energy and the Wealth of Nations – An Introduction to Biophysical Economics*. I draw quite frequently and without apology from this key resource.

Steve Bartell
Highwood Farm,
June, 2024

About the Author

Steven M. Bartell is a quantitative systems ecologist and modeler of complex ecological and environmental systems. He has applied his modeling skills and experience in assessing the impacts of diverse physical, chemical, biological, and ecological stressors on primarily aquatic resources, including endangered species, communities, food webs, and ecosystems. Bartell has variously addressed ecological risks posed by physical habitat degradation, excessive nutrient loadings, pesticides and other chemical contaminants, trace metals, ionizing radiation, invasive species, and climate change. He has modeled aquatic systems ranging in scale from 1-m^3 mesocosms, small streams, larger river systems, lakes and reservoirs, the Great Lakes, coastal marine systems, and the Gulf of Mexico.

Bartell has technically supported studies and projects in theoretical systems ecology, ecological risk estimation, natural resource damage assessments, and ecosystem restoration. His interests in the restoration and management of sustainable ecosystems led to initial explorations in the realm human dominated systems, particularly in relation to sustainable development, ecological economics, and biophysical economics. Bartell's discovery of Graham Turner's evaluation and basic confirmation of the 1972 Limits to Growth projections combined with a rather fortuitous opportunity to review drafts of the seminal work by Charles Hall and Kent Klitgaard, Energy and the Wealth of Nations, provided the major impetus for this initial foray into how a biophysical economy might eventuate.

Bartell currently serves as the Series Editor for the CRC Press, *Applied Ecology and Environmental Management,* largely in honor of the late Professor Sven Eric Jorgensen, who was the indomitable source of energy and enthusiasm in the production of most of the previous contributions to the series.

Acknowledgments

I have been fortunate to know and collaborate with a substantial number of "systems" thinkers and practitioners during my decades of professional participation – too many to thank individually, although many names come instantly to mind. I have undoubtedly been influenced through valued, collegial discourse and have further benefited from their contributions to the fields of systems analysis and quantitative modeling. I am eternally grateful for the inspiration and guidance provided by mentors Marty Meyer, Sumner Richman, Timothy F.H. Allen, and Jim Kitchell in the early years of my professional development.

I have been lucky in having the opportunity to work in academia, government institutions, and private corporations that similarly value theoretical and applied systems research, analysis, and assessment. I have been fortunate to participate in addressing a wide range of ecological and environmental challenges and these experiences in no small part stimulated my recent interests in complex, integrated, and adaptive socio-economic-ecological systems – hence this book, an initial foray into human dominated, yet hopefully sustainable systems.

I express my appreciation to Irma Shagla Britton, Senior Editor, CRC Press – Taylor & Francis Group for her counsel, support, and patience throughout the preparation of this book. This book would also not have been possible without the technical guidance and tireless support provided by Chelsea Reeves, Senior Editorial Assistant, CRC Press – Taylor & Francis Group. I am indebted to you both.

I am grateful for the encouragement and support provided by my wife, Cindy, who endured the development of the book and selflessly took on extra duties to help free up time for me to devote to preparing the manuscript.

Finally, any misinterpretation or misrepresentation of works described in this book are entirely my own and I apologize to the original authors as warranted.

Steve Bartell
Greenback, Tennessee

Section 1

Introduction

1

Introduction

1.1 This Book

Humans inhabiting Earth require a different dominant economic paradigm if they are to survive and flourish (Hall and Klitgaard 2018). The development of such an alternative will require sound underlying theory needed to guide the design and implementation of the necessary economic structures and processes conducive to sustainability, perhaps eudaimonia. Dramatic alterations from current economic practices come with obvious challenges – economic, social, political, and ecological. The book will identify and address many of them and offer a potential solution in the form of a biophysical-based economy as an alternative to the current neoclassical economic paradigm.

The book is structured as three broad topic areas each supported by several chapters. The introductory Section 1 includes three chapters. Chapter 1 reviews the fundamental concepts of an economy and revisits the current dominant ecological paradigm – namely neoclassical economics and its emphasis on continued growth. The chapter briefly summarizes different economic systems that have by and large come and gone. The intent is not to give an in-depth analysis of historical economic activity, but rather to provide context for suggested changes needed to establish a self-sustaining economy on a finite planet. The initial chapter concludes with a discussion of existential risk (human extinction) as an all too possible – perhaps inevitable – outcome of the neoclassical model. Existential risk emerges as a prime motivator for the quest for an alternative economic system and sets the stages for the remainder of the book.

1.2 Neoclassical Economic Paradigm

The proposition underlying this book as that the contemporary neoclassical economic model needs to be transformed into a model that is compatible with the finite nature of Earth's physical-chemical-biological (ecological)

DOI: 10.1201/9781003308416-2

processes, from which real wealth derives. The neoclassical model defines the economy as consisting of households and firms (Figure 1.1). The model is described simply and presented in this book more for the sake of convenience for readers (like ecologists and environmental scientists) who are not immersed in the study and lexicon of economics on a routine basis. The intent here is mainly to provide sufficient context to underscore the current dilemma of striving for a sustainable economy while continuing to embrace a fundamentally unsustainable economic paradigm. The presentation draws unabashedly from Hall and Klitgaard (2018) and the interested reader can find additional detail there.

In the neoclassical economy, goods and services flow from the firm to the households, which, in turn, provide land, labor, and capital to the firm. Correspondingly, money flows from the firm to the households in the form of wages, shared profits, and so forth. The members of the households, in turn, purchase from the firm. Participants are located either in households or in firms. Their only interactions are in the form of transactions. The neoclassical model describes self-maintaining equilibrium conditions where all of the goods and services produced by the firms are purchased by households, which, in turn, provide a balancing supply of land, labor, and capital to produce the goods and services. All participants spend their incomes on consumption or production. Households do not save; firms do not invest. There are no taxes. The government does not spend. There is no trade (Hall and Klitgaard 2018).

The neoclassical model is fundamentally transactional in nature. Money is presumed to cycle indefinitely in relation to the production of goods and

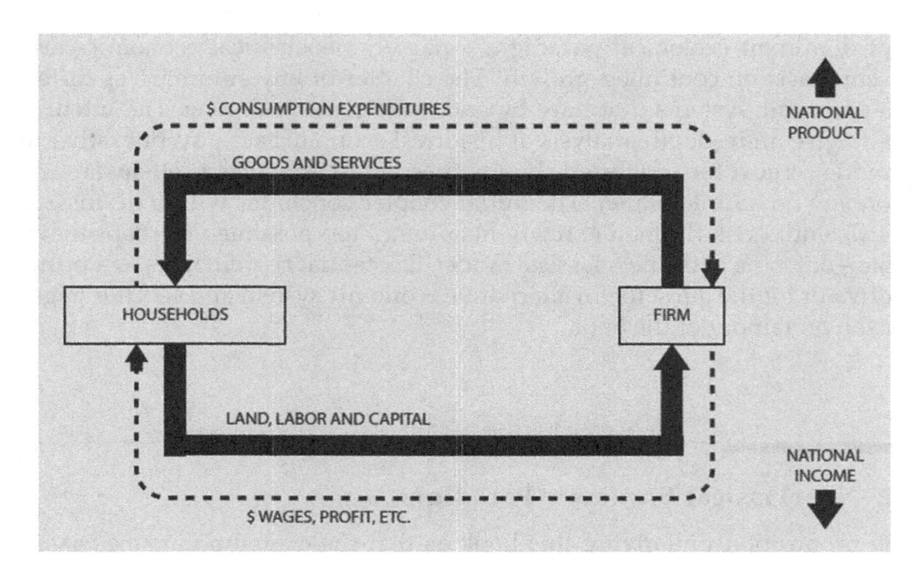

FIGURE 1.1
The neoclassical economic model (from Hall and Klitgaard 2018).

services. Production (Q) is defined solely as a function of labor (L) and capital (K) according to

$$Q = Ak^{\alpha} L^{\beta}$$

Where α defines capital's share of output and β defines labor's share. Additionally, $\alpha + \beta = 1$. The constant A multiplies capital and labor and roughly represents "technology" (Hall and Klitgaard 2018).

Central to this model based on exchange is the concept of the market as a place for exchange and trade. The basic idea was that individuals would seek to maximize their well-being within the limits of resources at their disposal. In an ideal ("free") market, consumers will buy goods and services to satisfy their needs. Producers of the goods and services will shift their focus to producing those items that customers want – a situation which will generate the largest profits for the producers and maximum benefits for the consumer. Interactions between supply and demand for goods and services define equilibrium price and quantity.

The circular economic model is attractive in theory because it implies an economy that is self-contained and self-regulating. Being self-contained means, to economists, that the economy is a proper subset of itself. This perspective implies that human interactions are merely economic transactions. Importantly, the real world lies external to the self-contained circular economy. A key challenge in building a biophysical economy lies in introducing real-world function into the theoretical equilibrium model underlying current economic theory and practice. A good portion of the book addresses this challenge.

The self-regulated economy implied by Figure 1.1 is assumed to produce outcomes that are efficient and equitable (Hall and Klitgaard 2018). By efficient is meant that, left to its own devices, the circular economy will produce allocations of resources that maximize well-being for all participants. In an equitable economy, individuals become better off in direct relation to their contribution to productivity. Mainstream economists further assume that these attributes are time invariant – the relations between humans and nature are essentially the same today as during medieval or ancient times. The future will remain like the present in terms of the structure and function of the basic circular economic model. The substantial inequalities in the distribution of wealth in the current economy call into question the implied efficiency in maximizing well-being for all.

1.3 Economic Growth

Conversations about sustainability and biophysical economics are, in the end, largely about economic growth – or to the point, diametrically opposed perceptions of the real-world implications of continued growth (Verstegen and Hanekamp 2005). Economic growth can be defined as an increase in capacity to supply ever-diversifying goods and services to a population. The increased capacity is believed to be based on advancing technology and corresponding institutional and ideological adjustments (Kuznets 1973). Romer (1996) would contend that a combination of abundant resources and large markets, in addition to technology, for example, the early-19th-century United States, were necessary to grow, and, importantly, to grow in scale. To this mix, new ideas and increases in "useful knowledge" were also viewed as critical to the manifestation of significant economic growth, as characterized by the Industrial Revolution (Mokyr 2007, 2005).

Prior to the Industrial Revolution, economies did not grow substantially over long periods of time (Hall and Klitgaard 2018). The more recent historical emphasis on economic growth appears to have accelerated in the 1930s in relation to mass unemployment and extreme poverty (Verstegen and Hanekamp 2005). To help characterize economic growth in more tangible terms, Kuznets derived the gross domestic product (GDP) as a measure of the overall value of goods and services produced by a country in a year, where GDP = consumption + government spending + private investments + exports – imports (Schwab 2021). Kuznets allegedly cautioned against using the GDP as the basis for economic planning. This caution has been obviously ignored by mainstream economists and planners. Following World War II, economic growth continued to raise standards of living and achieve full employment. Growth in absolute terms (i.e., GDP per capita) has been lauded additionally as a source of income for private and public investments in technology, health care, education, social welfare programs, and even environmental protection (Verstegen and Hanekamp 2005). The absence of growth (recession) or a deliberate policy of zero growth implies corresponding increased unemployment, reduced standards of living, and increased economic inequality. The remedy to avoid such a downward spiral in economic conditions is believed to be never-ending growth – which is entirely possible, even necessary, assuming the currently dominant economic model (Figure 1.1). This model appears entirely feasible in a world defined by infinite resources and infinite capacity to assimilate the wastes and other insults associated with unbridled growth. Planet Earth is unfortunately not that world (e.g., Klitgaard and Hall 2018; Kitzes and Wackernagel 2009; Meadows et al. 1972 – and others).

1.4 Dynamic Carrying Capacity

The current exploration into the creation of an economy based on biophysics is not inherently anti-growth. But sustainable material growth in absolute terms on a finite Earth is inarguably an oxymoron. The underlying premise is that a finite planet imposes limits to indefinite economic growth. This limit can be discussed in terms of carrying capacity – a fundamental ecological concept (e.g., Odum 1971). Every species, including *Homo sapiens*, experiences limitations to organic (exponential) growth – the corresponding population size (numbers, biomass) defines the carrying capacity. A key observation is that carrying capacity is not a static value, but it can change in time and/or space according to changes in the availability of one or more resources that currently constrain further growth – that is, carrying capacity is dynamic. The concept dates at least to Malthus and likely earlier.

Importantly, *H. sapiens* demonstrates an ability to measurably influence its dynamic carrying capacity. Most species do not purposefully modify or enhance their carrying capacity. Their populations increase and decrease (feast and famine, boom and bust) in relation to changes in resource availability determined by biogeochemical processes not in their direct control – equivalent to an ecological "luck of the draw." In contrast, humans have developed technologies and understanding that permit management of at least renewable resources (e.g., agriculture, livestock, timber, aquaculture, fisheries) and energy (e.g., solar, wind, hydropower – nuclear?) that contribute to economic well-being (and population size). Admittedly, capabilities for managing renewable energy sources and resources are variously imperfect and evolving. Yet a global population of some 8 billion (at the time of writing) attests to the success of humans in managing their ecological context and increasing carrying capacity from that of early hunters and gatherers. Humans have so far proven adept at expanding their ecological niche through cultural niche construction (Waring et al. 2023).

1.5 Existential Risk

A key motivation for insisting (at least) on a conversation of how a biophysical economy might be constructed and implemented derives from the inescapable recognition that humans and their economy are energetically open, dissipative systems (Odum 1971; Hall and Klitgaard 2018). These systems are not physically represented to any useful extent in the circular household-firm economic model (Figure 1.1). The implications of continued economic growth on human persistence in a more realistic description of the complex interrelated ecological-economic systems were projected more than 50 years ago (Meadows et al. 1972). The "business-as-usual" model results indicated major global economic impacts and corresponding dramatic reduction in human population size in the timeframe of 2020–2040 – that is, now. The more recent work of Turner (2008, 2014) uses 30+ years of data to evaluate the performance of the 1972 *Limits to Growth* projections and concludes that the world is tracking the "business-as-usual" model results with scary accuracy.

Independently, Bostrom (2013, 2009, 2002) and others have suggested a 1-in-3 risk of human extinction before 2100 as resulting from a combination of factors that derive from the unsustainable ways humans inhabit the planet. Importantly, Bostrom and colleagues at the Oxford Institute for Humanity define existential risk (human extinction), not as zero *H. sapiens*, but as significant reductions in population size, loss of technology, and critically, the loss of the ability to re-create technology. The corresponding devolution of institutions and order that define society plunge survivors into a "Mad Max" world scenario.

An economy based on biophysics is offered as one possible remedy to "soften the landing" in relation to the *Limits to Growth* model projections and reduce existential risk from human-mediated sources. The premise is that by operationally connecting economic activity (growth) to the renewal rates of key natural resources, managed and unmanaged, humans might sustain and perhaps even flourish on Earth indefinitely. The challenge is wicked in the sense that a solution requires operating (growing) within planetary constraints, which implies a planned economy – at least planned to some extent. Planned economies demonstrate historically poor track records. Planning must be sufficient in scale and dimensionality to maintain the system within safe operating conditions (e.g., Steffen et al. 2015; Rockstrom et al. 2009), but not so overbearing as to stultify initiative, enthusiasm, or freedom for creation and exploration of alternative socioeconomic-ecological systems (Giampietro and Mayumi 2018; Verstegen and Hanekamp 2005).

An important theme underscoring the presentation is that the structures and processes generally afforded to the branch of study termed "economics" are human inventions and can be changed by humans. In contrast, the biophysical structures and processes which simultaneously serve as the source of real wealth and impose constraints on growth are not particularly negotiable.

1.6 Brief Chapter Introductions

The book accordingly focuses on the following key questions:

What is biological economics? Answering this question will draw on substantial descriptions and justifications developed over decades of discussion (e.g., Hall and Klitgaard 2018; others). The fundamental concept that defines biophysical economics is the recognition that real wealth and economic sustainability require an economic model that is compatible with the biophysical life support systems that simultaneously permit and ultimately constrain economic activity on Earth. This model contrasts with the current neoclassical paradigm that requires continuous growth in absolute terms – with corresponding benefits to many and increasing existential risk for all.

How might a working biophysical economy be constructed? Addressing this question emerges as a central theme of the volume. Much effort has been devoted to the need for and justification of a biophysical economy. By comparison, scant attention has focused on how a biophysical economy might be designed and put in place. Additionally, if such an economy were to be implemented, how would we know? The basic premise underlying the discussion in this book is that a sustainable economy based on biophysics can be implemented through modifications – some relatively minor, some severe – of existing economic and political institutions, structures, and processes. Wholesale reinvention of the current economic system is believed unnecessary. The purpose is to at least begin the conversation, recognizing that drastic changes are least likely to be undertaken – but some changes will be uncomfortable challenges to the economic status quo.

The Universe is largely and for the most part deadly hostile to life as we know it. Chapter 2 addresses the fragility of carbon life forms and their "niche space" on Earth as a prime motivator for establishing an economic system that respects the universally unlikely condition of life as we find it on this planet. The chapter reminds us how unique the planet is in terms of conditions favorable to the evolution and sustenance of life as we know and measure it.

Chapter 3 further develops the need for a sustainable economic system that is compatible with the finite nature of the planet. This chapter addresses economics within the context of planetary boundaries – biophysical aspects of a functioning planet that must remain intact in order for humans to persist indefinitely. The boundaries are further discussed in terms of the projected limits to growth introduced in the 1970s by Meadows and colleagues. The *Limits to Growth* (1972) forecasts of the potential for global economic collapse are revisited with the advantage of having several decades of observations along multiple dimensions to evaluate model performance so far. Chapter 3 ends by introducing the concept of finite singularity – a mathematical consequence of dynamic systems growing faster than exponential (West 2017). One key observation is that human population growth continues to accelerate – doubling times have been decreasing, which implies the possibility of realizing a finite singularity, where the demand for at least one necessary resource will increase exponentially in real time – a clearly unsustainable condition, a breaker of one or more planetary boundaries, and a harbinger of human extinction (Bostrom 2013, 2009, 2002).

Section 2 continues the conversation by focusing additional attention on what might be foundational to the implementation of a biophysical economy. The design and implementation of an alternative economic system compatible with the planetary life-support systems – that is, a biophysical economy – are introduced in Chapter 4. The central theme of the book underscores replacing the current economic system with one based on biophysics.

Correspondingly, Chapter 4 briefly reviews the history of biophysical economics and outlines its key attributes. Chapter 5 begins the conversation concerning economic transformation with an initial suggestion of what a biophysical economy might look like – how it might be structured through modification of existing economic structure and processes. Chapter 5 concludes with a discussion of the meaning of these peculiar life-supporting conditions in the context of managing renewable and non-renewable resources in a manner consistent with the persistence of humans and ecosystems. In a sense, Chapter 5 presents the entire book in short form.

Chapters 6–10 constitute the remainder of Section 2 and focus on current pillars of a biophysical economy and recognize the importance and necessary contributions from the private sector, the public sector, government, finance, and policy needed to make such a dramatic shift from the neoclassical paradigm to a sustainable, biophysical economic model. A fundamental premise underlying the development is that the likelihood of actually instantiating an economy consistent with the fundamental attributes of biophysics will be inversely proportional to the degree to which existing institutions and processes will have to change. At the same time, it is recognized that any meaningful transition to an economic alternative to the current model will not proceed without discomfort. Chapter 6 briefly reviews capitalism and explores the reinvention of capitalism required to move towards a sustainable economy. The science of corporations is introduced as one possible avenue to explore to develop causal relationships among the structures and processes inherent to corporation and anticipated outcomes of management actions intended to increase corporate sustainability and sustainability at larger scales as a result of wholesale adoption of sustainable corporate practices. Current corporate activities in the dimensions of environment (E), social (S), and governance (G) are addressed from the perspective of the potential contributions of corporate ESG analysis, corporate function, and corporate contributions to a sustainable biophysical economy.

Chapter 7 explores fundamental changes in the public sector (that is, the general public) required to put into place an operational economy based on biophysics. The chapter asks whether *Homo sapiens* is sufficiently evolved to the extent of being capable of acting on its own behalf to thwart the existential risk implied by continuous growth and finite singularity. The chapter concludes with some consideration of a new consumerism as one antidote to the neoclassical economy. People in the Anthropocene (really the "Capitalocene") are both their own economy and their own ecology – two sides of the same planetary coin.

Government institutions, as Chapter 8 describes, will necessarily play a pivotal role in any transition from the neoclassical paradigm to a sustainable economy grounded in biophysics and Earth's life support systems – the origins of real wealth. The chapter focuses on government structures and processes in the United States, but the concepts ought to extend to other countries as well. Sustainability, in the end, has real meaning and value only at a global scale. Contributions from the executive, legislative, and judicial branches of government, as well as supporting agencies, are briefly discussed from the vantage point of building an actual biophysical economy. The government collects and manages a wide variety of demographic, economic, energy, and environmental data at nation scale that could be used to help design a biophysical economy. A key challenge in implementing a biophysical economy lies in being able to analyze and cross-walk the data using technologies of "big data" and artificial intelligence to usefully measure economic performance and corresponding human health and ecosystem integrity across all the dimensions of sustainability.

Nothing happens at sufficient scale without corresponding finance. Chapter 9 explores the implications of a shift from the current economic paradigm to one couched in biophysics, life support, and sustainability on a finite planet. The neoclassical model is fundamentally a positive feedback loop – where growth begets growth in an endless cycle presumed inviolate into the indefinite future. Thermodynamics cautions us otherwise. The chapter examines how current financial institutions might help facilitate the transition to a biophysical paradigm. Implications of slower (or no) growth on equities (stock market) are considered, given their importance to the current financial well-being of millions of people.

Chapter 10 discusses a biophysical economy from a policy perspective. A biophysical economy is a planned economy and planned economies do not demonstrate an encouraging historical track record – not necessarily because of bad plans, more often from ineffective, often corrupt implementation. Creating a sustainable economy firmly grounded in biophysics will require significant policy decisions at scales ranging from local to global.

Section 3 aims to pull all the pieces together and ultimately prescribes a path to a sustainable economy consistent with planetary biophysics. Chapter 11 discusses the role of feedback mechanisms – positive and negative – in setting up an operational biophysical economy. Feedback mechanisms are fundamental in determining the dynamics of complex adaptive systems comprised of many interconnected components. The economy is certainly a complex system operating under far from equilibrium conditions.

Chapter 12 underscores the importance of causal models in exploring the efficacy of different approaches to designing and implementing a biophysical economy. Models that operate at global, regional, and local scales – down to the scale of individual corporations will likely contribute to putting a biophysical economy in place. Chapter 12 also introduces the concept of a "science of corporations" and explores its relevance in the evaluation of corporate performance in building a sustainable economy.

Finally, Chapter 13 aims to make use of the previous chapters to provide the context and guide the development of a "beta-version" of a biophysical economy. The end game is an initial prescription of what a biophysical economy might look like and a discussion of how it might operate.

Not by way of apology, but the sheer number of topic areas and socioeconomic and ecological dimensions that need to be at least mentioned in musing about the possibility of an operational biophysical economy preordained shallow and uneven treatments across the topic areas absent a lifetime devoted to the production of this volume. Again, the intention of this presentation is to initiate the conversation or explore if the conversation is even possible. Is it possible to build the kind of economy described and justified by the biophysical economists (e.g., Hall and Klitgaard 2018)? The human, social, environmental, and political implications of the "business as usual" neoclassical model are increasingly evident in our daily life experience. The current economy has undeniably produced great wealth for few, and it has raised the standard of living for many, while imposing substantial degradation on planetary life-support systems. An alternative economic model to continuous growth is unavoidable – is the biophysical economy a viable alternative? Can such an economy be constructed voluntarily or will human carrying capacity be imposed by forces beyond human control? Based on the information provided in this volume, the reader may draw his or her own conclusion and hopefully extend the conversation. But extending the conversation, while important, is not sufficient. The conversation must be carried to the points of power in the current system where the required nation-scale economic transition can become real-world policy.

2

Earth Systems

2.1 Fragility of Life on Earth

The physical forces of the Universe – even our own planet – are staggeringly beyond comprehension. For example, as I sit seemingly at rest and write this book at a latitude of 37.76 °N, I am moving at approximately 800 mph around the Earth's axis. If the planet suddenly stopped, I would be launched eastward at 800 mph with obliterating consequences. I am simultaneously moving around the Sun at about 67,000 mph – or 19 miles per second. Videos of astronauts working on the space station appear to show them nearly motionless, while they and the station itself whizzes around the planet at about 17,500 mph!

Quite by accident in the ever-evolving and expanding Universe, there are several important physical and chemical features of planet Earth that are conducive to carbon-based life forms. The following might appear trite to readers well versed in the environmental sciences, but the expectation is that mainstream economists who pick up this book might have given less thought to planetary life-support systems that are not explicitly included in the neoclassical economic paradigm. So the intent is to provide some more or less tangible examples of circumstances permissive of life on this planet:

> Earth is situated in the solar system such that temperatures and pressures commonly encountered on the planet permit water to exist mainly in its liquid state. Organisms, including humans, have correspondingly evolved as essentially organized volumes of water defined and constrained by cellular membranes.
>
> The Van Allen radiation belts, integral components of the Earth's magnetosphere, trap high-energy particles that might otherwise impact the planet's atmosphere and its inhabitants. The planet's molten iron core creates a magnetic field – a magnetosphere – that helps buffer Earth from damaging intense solar radiation and high-energy particles. This magnetic field keeps the atmosphere from being "blown away" by the constant stream of charged particles (plasma, solar wind) emanating from the Sun.

DOI: 10.1201/9781003308416-3

> The ozone layer protects carbon life forms from deadly ultraviolet radiation. Absent this protective layer, organic life would be at increased and largely unavoidable risk of incineration by UV radiation.

Large-scale patterns of atmospheric circulation distribute atmospheric heat around the planet in a reasonably predictable manner (Maher et al. 2019; Schneider 2006). Surface winds tend to be easterly (westward) at low latitudes, westerly (eastward) at mid latitudes, and easterly or nearly absent at higher latitudes (Figure 2.1). The rotation of the Earth dictates the direction of the mean surface winds. Differential heating of air at the equator causes air to rise and flow poleward (Schneider 2006). The economic significance of the surface wind directions have been evident for centuries as the trade winds that facilitated historical maritime commerce.

Similarly, global ocean circulation conveys warmer temperature water from equatorial regions to northern latitudes where it cools and sinks (Broecker 2002). Thermohaline circulation driven by changes in temperature and salinity define a global conveyor belt for the large-scale transport of ocean water (Figure 2.2). Recent concerns focus on the disruption of historical patterns of ocean transport in the North Atlantic associated with increases in ocean temperatures (Dima and Lohmann 2010). The corresponding implications for general climate conditions in continental Europe have become of increasing concern.

The characterization of large-scale and general patterns of atmospheric and ocean circulation are fundamental to describing and understanding climate change in a general sense. Regional and more local scale articulation of these large-scale phenomena through statistical analysis of climate and meteorological data, including regional downscaling (Doury et al. 2022;

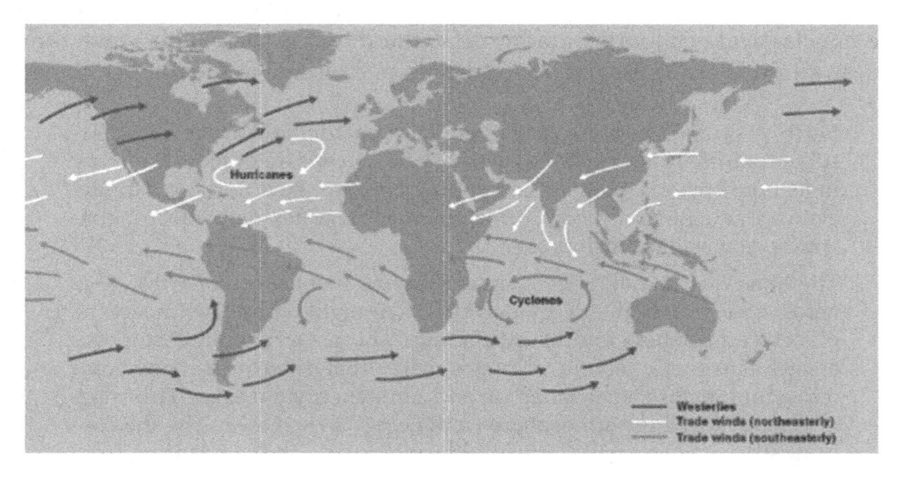

FIGURE 2.1
Illustration of global trade winds (adapted from NOAA).

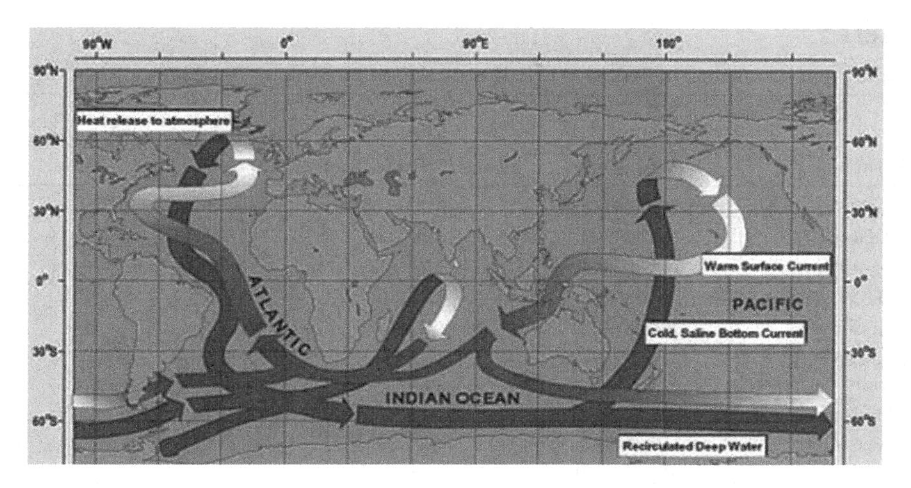

FIGURE 2.2
Large-scale ocean thermohaline circulation (from Woods Hole).

Walton et al. 2020; Wigley 2004) as well as sophisticated regional climate modeling (Giorgi 2019; Mearns et al. 1999), contribute to projections of climate change at scales important to economic planning and are therefore relevant to an initial conversation concerning the design and implementation of a biophysical economy,

2.1.1 Hydrological Cycle

Earth is further unique in that water exists in all three possible phases, water vapor, liquid water, and solid ice (Oki 2005). While the energy cycle on the planet is open and driven by solar radiation, the water cycle is comparably closed with the total amount of water on Earth remaining fairly constant on timescales less than geological. The hydrological cycle is driven in part by atmospheric circulation driven by unequal heating the planet's surface and atmosphere in relation to latitude (Oki 2005).

Table 2.1 lists the amounts of water in various resources based on summarized data from Oki (2005). Of course, the oceans comprise the predominant volume, followed by water in glaciers and groundwater. Lake water constitutes another key planetary water resource in addition to water in soils and in the atmosphere. Mean residence time of the various water resources range on the order of 10,000 years for ice in zones of permafrost to a few hours for biological water. The ocean mean residence time is about 2,500 years. Groundwater turns over about every 1,400 years. Lake water has a mean residence time of about 17 years. Atmospheric water turns over about every eight days, while soil moisture exhibits about a one-year residence time (Table 2.1). The scale and residence times of these various planetary sources

TABLE 2.1

Simplified Summary of World Water Reserves (adapted from Oki 2005)

Form	Area (km²)	Total Volume (km³)	Share (%)	Mean Residence Time
World ocean	361,300,000	1,338,000,000	96.54	2,500 y
Glaciers	16,227,500	24,064,100	1.74	1,600 y
Groundwater	134,800,000	23,400,000	1.68	1,400 y
Lake water	2,058,700	176,400	0.013	17 y
Soil moisture	82,000,000	16,500	0.0012	1 y
Atmospheric water	510,000,000	12,900	0.0009	8 days
Marsh water	2,682,600	11,470	0.0008	5 years
River water	148,800,000	2,120	0.0002	16 days
Biological water	510,000,000	1,120	0.0001	few hours
Total reserves	510,000,000	1,385,984,610		

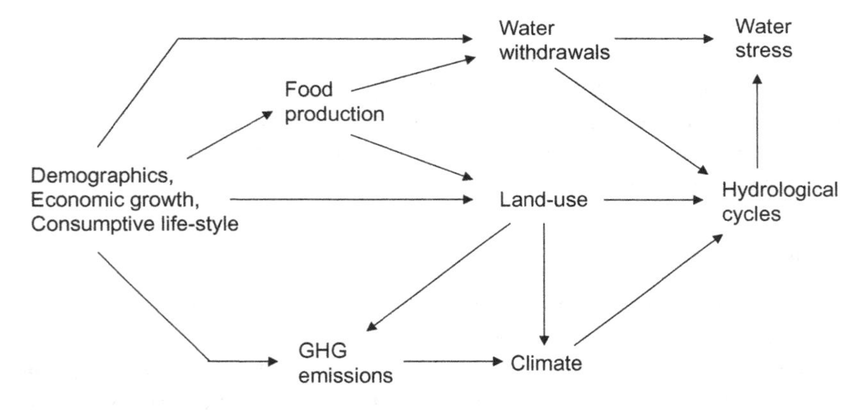

FIGURE 2.3
Major pathways and processes where demographic and economic growth impact hydrological cycles (adapted from Oki 2005).

of water are important to understand in the design of a sustainable economy that will invariably depend on reliable and sufficient water resources.

Human survival, as well as economic growth and prosperity are tied closely to the hydrological cycle (Figure 2.3). Understanding the consequences of economic activities that withdraw water for food production and industrial use in addition to municipal and domestic demands will be necessary for managing water resources in relation to economic sustainability, including the design and implementation of a biophysical economy. Water consumption fueled by demographic and economic growth can impact food production, lead to alterations in land use, and correspondingly exert negative impacts on climate and hydrological cycles (Oki 2005).

2.1.2 Nitrogen Cycle

Another tangible example of a planetary life support system is the cycling of nitrogen through its various forms and functions (Canfield et al. 2010). Nitrogen is the fifth most abundant element in the solar system and is essential for the synthesis of nucleic acids and proteins, which are fundamental to life. The atmosphere is dominated by N_2, but this gas is virtually inert. The associated biogeochemistry depends substantially on the reduction-oxidation reactions mediated by microorganisms. Inorganic nitrogen is necessary to support plant growth – both in terrestrial and in aquatic systems. Nitrogen fixation largely by bacteria convert N2 to ammonium (Stein and Klotz 2016). Nitrification also mediated by microorganisms oxidizes ammonium to nitrite and subsequently to nitrate, a form readily used by plants for growth. Dentrification defines a process of anaerobic respiration of nitrite to nitrous oxide and eventually to N2 gas to complete the nitrogen cycle (Table 2.2).

Importantly, the advent of the Haber-Bosch process in 1909 provided an industrial fixation of N2 to ammonia and provided synthetic N fertilizer that has come to play a key role in modern industrial agriculture (Stein and Klotz 2016; Canfield et al. 2010). Approximately 45% of the total annual nitrogen fixation on Earth is now anthropogenic (Canfield et al. 2010). Continued increases in the use of synthetic fertilizer has strained the nitrogen cycle to the point of excess N loading (runoff), eutrophication, and creation of "dead zones" (depleted oxygen) in coastal marine systems. The synthesis of urea from fossil fuels has also tied the cost of fertilizers to the price of oil. As post-peak oil manifests and the price of oil correspondingly increases, the associated increase in fertilizer cost and reliability of fertilizer supply will have to be addressed in the design and operation of a sustainable economy.

Even though the blue planet appears invitingly inhabitable from space, carbon-based life forms are constrained to a very small portion of Earth's totality. The atmosphere that supports us is extremely thin (~8 km) in comparison to the radius of the planet (6,400 km). As we increase our location from sea level to higher altitudes, we experience extreme conditions of frigid temperatures and decreased oxygen. As we descend into the depths of the ocean, atmospheric pressure and the weight of water increases to life-crushing

TABLE 2.2

Basic Processes Involved in the Nitrogen Cycle (after Stein and Klotz 2016)

Process	Description
Nitrogen fixation	Process whereby molecular N_2 in the atmosphere is converted to ammonium by bacteria that possess a nitrogenase enzyme
Nitrification	The sequential oxidation of ammonia to nitrite and nitrate
Denitrification	The microbial-mediated process where nitrate and nitrite are reduced to nitrous oxide (N_2O) and nitrogen gas, N_2

levels. The overall physical "niche" for organic life on Earth is remarkably small in relation to the overall size of the planet.

These overly simplified descriptions of life-critical phenomena are intended to remind us that we ought not (indeed, cannot) take life as we have come to know it for granted. These planetary limitations on life support are value-neutral, apolitical, and non-negotiable. The inheritance of culture and evolution of modern technology has permitted *Homo sapiens* to venture into otherwise unwelcome niche space – a process called niche construction (Waring et al. 2023). However, any economic system that fails to recognize the unlikely and fragile nature of life on Earth is not well positioned to meaningfully move towards sustainability.

2.2 Ecosystems and Life Support

The planet's life support systems can be described as ecosystem service functions provided by a broadly scaled collection of ecological and environmental biomes or ecoregions (e.g., Omernik 2004; Wright et al. 1998). Biomes comprise variously defined terrestrial and aquatic ecosystems whose resident populations and communities of organisms provide critical biogeochemical services that support life on Earth.

Daily (2000) offered a framework for managing life-support systems developed within the context of identified ecosystem services (Table 2.3). These services included the production of goods (food, fiber), regenerative processes, stabilizing processes, life-fulfilling functions, and conservation options. Daily (2000) asserts that proper management of ecosystem services could produce a reliable source of life support indefinitely.

At the same time, there are difficulties in quantifying direct links between ecosystem processes and human well-being (Harte 2001) because ecological services are

- often invisible. Harte (2001) provides an example of denitrifying bacteria that contribute nitrous oxide through their metabolism. The nitrous oxide moves to the stratosphere where it contributes to the effectiveness of ozone in protecting organisms from ultraviolet light.
- complex. Harte (2001) offers the challenges in maintaining genetic diversity in populations of direct commercial importance (e.g., wild salmon).
- probabilistic and spreads over space and time. Degradation of human well-being is difficult to relate directly to degradation of ecosystems. Consider the impacts of massive deforestation on downwind increases in the likelihood of drought (Harte 2001).

TABLE 2.3

Classification of Ecosystem Services with Examples (from Daily 2000)

Classification of Ecosystem Services
Production of Goods
Food
Terrestrial animal and plant products
Forage
Seafood
Spices
Pharmaceuticals
Medicinal products
Precursors to synthetic pharmaceuticals
Durable materials
Natural fiber
Timber
Energy
Biomass fuels
Hydropower
Industrial products
Waxes, oils, dyes, latex, rubber, others
Precursors to many synthetic products
Genetic resources
Intermediate goods that enhance other goods
Regenerative Processes
Cycling and filtration processes
Detoxification and decomposition of wastes
Generation and renewal of soil fertility
Purification of air and water
Translocation processes
Dispersal of seeds for revegetation
Pollination of crops and natural vegetation
Stabilizing Processes
Coastal and river channel stability
Species compensation under varying conditions
Control of pest species
Moderation of weather extremes (e.g., temperature and wind)
Partial stabilization of climate
Regulation of hydrological cycle (mitigation of floods and drought)
Life-fulfilling Functions
Aesthetic beauty
Cultural, intellectual, and spiritual inspiration
Existence value
Scientific discovery
Serenity
Preservation of Options
Maintenance of ecological components and systems needed to continuously supply goods and services

- rarely substitutable. Or substitutable at great cost to society. Harte (2001) further offers the example of the expensive use of pesticides required when insect-eating bird populations are decimated through habitat loss.

2.3 Sensitive Ecosystems

Nilsson and Grelsson (1995) define the fragility of ecosystems as the inverse of ecosystem stability and emphasize the potential for inhabiting species or communities to be strongly damaged by human activities. Fragility is characterized as an inherent property of ecosystems, but it remains difficult to quantify *a priori* as any anticipated response to disturbance.

2.4 Renewable Resources and Renewal Rates

The biophysical guiding principle regarding the utilization and management of renewable resources is remarkably simple in concept: rates of utilization must not exceed rates of renewal (natural or managed) over economically relevant scales in time and space. Implementing this simple guiding principle is not simple.

Rates of renewal for unmanaged resources can vary substantially in space and time. Developing an accurate and reliable estimate of resource renewal to guide utilization will have to accommodate such variability and uncertainty. The renewal rates of managed resources might be expected to show less variability in relation to the success of management. One important concept concerning the management of renewable resources (e.g., fisheries) is maximum sustainable yield (Ulrich et al. 2017). This approach attempts to take advantage of comparatively high rates of production for populations in a logarithmic growth phase. Harvesting is limited to values that sustain the managed population in a phase of rapid regrowth. More recent adaptation of the maximum sustainable yield approach has extended management focus beyond the population of interest to include a more comprehensive ecosystem context – that is, ecosystem management, where ecological services and biological resources are conserved while sustaining human uses of these resources (DeFries and Nagendra 2017; Zabel et al. 2003; Brussard et al. 1998).

Brussard et al. (1998) identify seven key components of ecosystem management that include

1. delineating or defining the ecosystem to be managed,
2. articulate the strategic management goals,

3. develop a comprehensive characterization of the selected ecosystem,
4. obtain corresponding socioeconomic data,
5. develop an integrated model linking the ecological and socioeconomic data,
6. implement selected management actions,
7. monitor the results of management to assess success or failure.

These component steps can be readily transformed into an adaptive management framework by linking the results of monitoring back to the earlier steps in the process where management goals might be revised or management actions might be modified to achieve the desired results or efficiently learn that the desired outcomes are not feasible given current management capabilities. Adaptive management – learning while doing – appears as a reasonable and prudent approach to the implementation of a biophysical economy given the complexity of the challenge, incomplete understanding of socioeconomic, political, and environmental systems, and the unproven ability to plan and manage an ecönomic transition at a nation-scale.

2.5 Managing Non-renewable Resources

The sustainable utilization of non-renewable (exhaustible) resources has been the subject of detailed analysis and modeling (e.g., Bastianoni et al. 2009; Martinet and Doyen 2007; Hotelling 1991; Stiglitz 1974). Prudence suggests using these resources as minimally as possible while striving to meet overall economic objectives (e.g., growth). The mathematical approaches describe the optimal allocation of an exhaustible resource over time through the use of utility functions that are maximized. A key result is that the stock of the resource and its optimal consumption tend towards zero over time. The neoclassical assumption is that increasing demand for a dwindling resource (e.g., oil) will stimulate finding or inventing an alternative or substitute for the resource of concern. Innovation and technology will circumvent resource depletion, although substitutes for arable land, breathable air, and potable water are difficult to conceive of at scale.

The Universe appears largely hostile to organic life as we know it. True, scientists have estimated that there are large numbers of planets with life-supporting characteristics similar to Earth distributed throughout the vastness of space. Yet, this large number is a vanishingly small fraction of all likely planets. Earth as a life-support system appears as an astronomical anomaly – a low probability event with high consequences, at least for its inhabitants.

3

Planetary Boundaries

3.1 Boundaries

A fundamental tenet underlying the concept of a biophysical economy is that the physical, geological, hydrologic, chemical, biological, and ecological structures and processes measurable on Earth together, as integrated systems, both support and constrain human activity on the planet. An economy on Earth is necessarily an open thermodynamic system (Hall and Klitgaard 2018; Odum 1971). The natural laws of physics, chemistry, and biology are directly relevant to the design and implementation of a sustainable economy. These laws are apolitical and not particularly negotiable.

Rockstrom et al. (2009) further pursued the concept of integrated earth life-support systems ultimately constraining human activities within a sustainable range. The following largely paraphrases this seminal work. The authors introduced planetary boundaries as defining a multi-dimensional safe operating space. Transgressing planetary boundaries, individually or in combination, might prove deleterious or even catastrophic to the human (and non-human) enterprise. They speculated that crossing thresholds could result in non-linear and abrupt changes in environmental conditions at planetary scales. Rockstrom and colleagues defined thresholds as "nonlinear transitions in the functioning of coupled human and environmental systems" (citing Schellnhuber 2002; Lenton et al. 2008).

The development of planetary boundaries is based on three avenues of investigation (Rockstrom et al. 2009):

- The magnitude of human activities in the context of the capacity of Earth life-support systems to sustain such activity, or the biophysical constraints on economic expansion,
- Increased understanding of Earth system processes necessary to further global change research and the science underlying sustainability, and
- The integration of resilience, complex system dynamics, and self-regulation of living systems.

 DOI: 10.1201/9781003308416-4

TABLE 3.1

Proposed Planetary Boundaries (Adapted from Rockstrom et al. 2009)

Earth System Process	Control Variable	Planetary Boundary
Climate change	atmospheric CO_2, ppm	350–550 ppm
Ocean acidification	Carbonate ion concentration	Sustain >80% of pre-industrial aragonite saturation
Stratospheric ozone depletion	Stratospheric O_3 concentration, DU	<5% reduction from pre-industrial level of 290 DU
Atmospheric aerosol loading	Regional atmospheric particulate concentration	To be determined
Biogeochemical flows	P inflow to oceans compared to background; N_2 removed from atmosphere for human use, Mt N/y	P: <10x (10x–100x) N: 35 Mt N/y or about 25% of annual N fixation by terrestrial ecosystems
Global freshwater use	Blue water consumed, km^3/y	<4,000 km^3 (4,000–6,000 km^3)
Land-system change	Percentage of global land converted to cropland	≤15% of ice-free land surface (15–20%)
Biodiversity loss	Extinctions/million species/y or (E/MSY)	<10 E/MSY (10–100 E/MSY)
Chemical pollution	Concentrations or effects on ecosystems or Earth system function	To be determined.

The overall approach incorporates previous work on limits to growth (Meadows et al. 1972), safe operating standards, the precautionary principle, and tolerable ranges. Following their approach, Rockstrom et al. (2009) identified nine planetary boundaries (Table 3.1).

3.2 Planetary Boundaries Status

Steffen et al. (2015) continued the development of the concept of planetary boundaries and provided an update on the status of individual boundaries (Figure 3.1).

The results of the Steffan et al. (2015) analysis indicate that losses in genetic diversity and the biogeochemical flows of phosphorus and nitrogen present high risks for breaking planetary boundaries. Climate change and changes in land use suggest increasing risk of violating planetary boundaries. Results of the recent IPCC analysis of climate change (Crimmins et al. 2023) would support an increased risk associated with climate change. As this volume

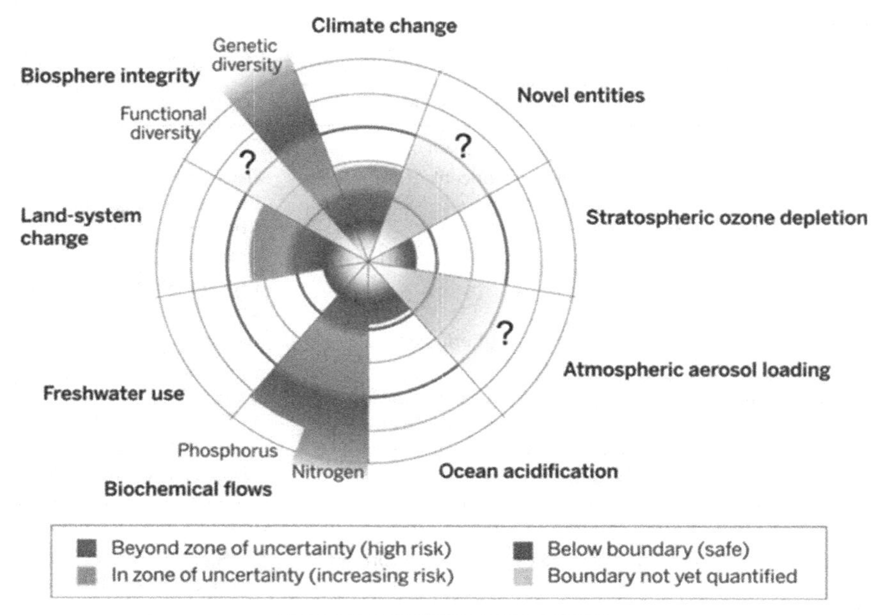

FIGURE 3.1
Estimates of current Earth system conditions in relation to selected planetary boundaries (from Steffen et al. 2015).

goes to press, global carbon emissions continue to increase, fossil fuels drive energy production, and global average temperature is approaching and, in several instances, exceeding the 1.5 C threshold.

Of particular relevance to biophysical economics, Rockstrom et al. (2009) recognize and emphasize that

> "Thresholds in key Earth System processes exist irrespective of peoples' preferences, values, or compromises based on political and socio-economic feasibility, such as expectations of technological breakthroughs and fluctuations in economic growth."

Planetary boundaries are directly relevant to the development and implementation of a biophysical economy. The boundary values or thresholds could serve as key benchmarks in evaluating the performance of economic activity aimed at achieving a sustainable economy in relation to the planet's biophysical support systems. Metrics that usefully quantify each of the boundaries should be developed and used as endpoints in models and analysis aimed at evaluating the longer-term sustainability of the current economic paradigm and in the design and implementation of alternatives, including a biophysical economy.

3.3 Breaking Ecological Constraints

One of the major challenges in sustaining economic activity on the planet is the fact that humans are simultaneously inside and outside of the planetary biophysical system. As carbon life forms, humans demonstrate the same requirements as other organisms in terms of breathable air, water in sufficient quantity and quality, physical habitat (shelter), and an energy source (food). Earth's evolving, but finite, life-support systems historically both provided these requirements and constrained human population size.

As humans developed technology and acquired artifacts, they were able to effectively break many of the previous ecological constraints – hence, they left much of their historical ecological context and moved outside the system (Waring et al. 2023). Breaking these constraints also contributed to substantial increases in population size to today's 8+ billion (and growing). Being outside the system, we have reached the point where humans and their artifacts have not only broken ecological constraints, but also now threaten the very existence of life on Earth – an existential risk (e.g., Bostrom 2013, 2009, 2002). As summarized by Costanza et al. (2007a), Diamond (2005) identified multiple interacting factors that have led previously to the collapse of historical societies, including:

- Population growth and levels of human consumption (e.g., overfishing)
- Availability of freshwater
- Loss of habitat, biodiversity, and ecosystem services
- Erosion and degradation of soils
- Limits on photosynthetic capacity and other available energy sources
- Invasive species, toxic chemicals, and climate change

One challenge in designing and building a biophysical-based economy lies in determining to what extent must we put humans "back inside" the ecosystem in order to sustain and flourish.

3.4 Limits to Growth

Of course, the idea that organisms, including humans, cannot expand indefinitely on a finite planet is not new (Malthus 1986, 1798). Finite resources imply that every species has a carrying capacity – *Homo. sapiens* is no exception. The question of how many people can be supported on Earth has been a subject of conjecture and debate at least for centuries (Cohen 2017, 1997, 1995).

Perhaps the classic works of Meadows and colleagues (Meadows et al. 2004, 1992, 1972) remain as the focal points for projecting the global implications of continuous growth. The "business as usual" model scenario presented in the original *Limits to Growth* (Meadows et al. 1972) indicated global collapse around 2030 with subsequent reductions in human population size (Figure 3.2). Other modeled scenarios suggested that collapse could be avoided by timely changes in policy, cultural behavior, and technology. Upon publication, the model projections, indeed the concept of limits to human growth, were summarily dismissed by mainstream economists (Turner 2008).

Even though the original Limits to Growth sold countless numbers of copies and the book was translated into 30 languages, the work was largely ignored in the decades following its 1972 publication. Mainstream neoclassical economics continued to dominate economic thought, policy, and practice.

A significant turn of events occurred when in a 2008 publication, Graham Turner compared the original "business as usual" model projections with 30 years of data collated since the original 1972 publication of *Limits to Growth*. The agreement between modeled projections and the data for the 1970–2000 period of comparison was compelling and correspondingly frightening (at least to this author). Based on the metrics selected and quantified by Turner (2008), the continuous growth model scenario was tracking the available data for the six key model output variables (Figure 3.2). One potentially saving observation is that none of the tracked per capita dimensions, population,

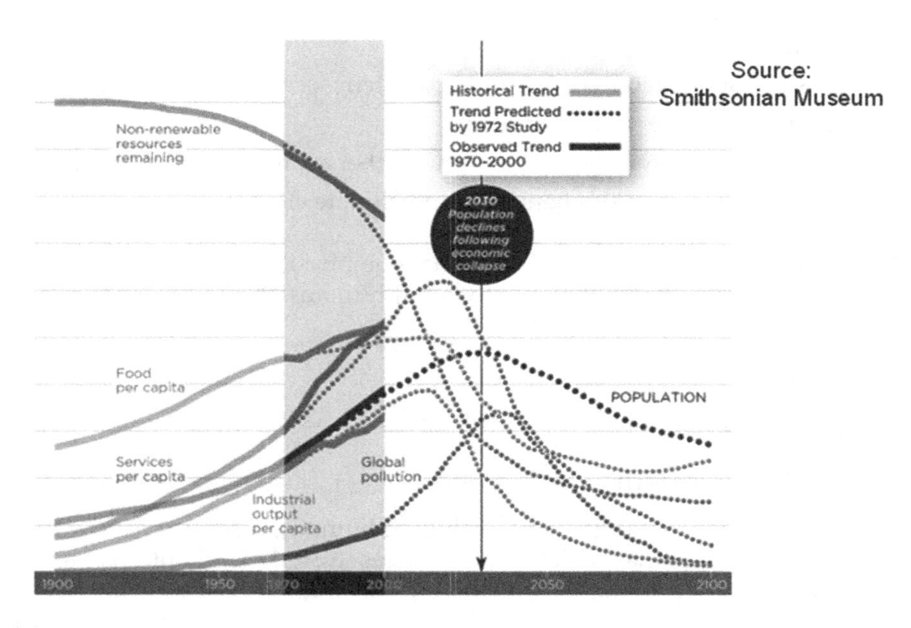

FIGURE 3.2
Limits to growth revisited (after Motesharrei et al. 2014).

or global pollution in the model projections have reached a period when the trajectories shift from increasing to decreasing. The model suggests population decreasing on or around 2030. Robust evaluation of the accuracy of the model would be enhanced if data describing reductions in these key model dimensions matched up with the predicted tipping points and subsequent downward trends. Time will tell, but the model:data agreement for these projections will clearly come too late to take effective action to avoid collapse. The ultimate "I told you so" (i.e., Meadows et al. 1972) would come at a major expense to the human enterprise.

Regardless of the assumptions and potential technical issues in the model:data comparison (Turner 2008), the overall results remain quite compelling. Subsequent incorporation of additional years of data into the comparison appeared to further bolster the agreement between model results and real-world observations (Turner 2014 – and see Herrington 2022). The agreement between the *Limits to Growth* model projections and available data was a key stimulus for exploring how a biophysical economy might be set up as a possible alternative to collapse. The analysis by Turner and colleagues is either correct or in error – regardless, the results of model:data comparisons to date seem to warrant serious consideration of alternatives to business as usual, if nothing other than to perhaps "soften the landing" should collapse be unavoidable. A biophysical economy might be argued as one such alternative to collapse (Hall and Klitgaard 2018).

3.5 Ecological Footprint

Another approach to characterizing the implications of current economic demands on key ecological resources and environmental support systems was introduced as the concept of an ecological footprint (Rees 1992; Wackernagel and Rees 1996; Wackernagel et al. 1999). In its simplest description, the ecological footprint measures the amount (ha) of Earth surface area and its attendant resources required to support a demand. Demand might be for an individual, groups of people (i.e., a nation), selected activities such as manufacturing, or even for the entire planet (Kitzes and Wackernagel 2009). The footprint measures the areas required to support demands for food, fiber, timber, energy, and space for infrastructure. Area is also required to absorb waste products.

Calculation of the footprint reflects current technologies (Kitzes and Wackernagel 2009). With increasing technological efficiency in some sectors (e.g., energy production, manufacturing), their contributions to the overall footprint have decreased over time. Importantly, there are thermodynamic limits to efficiency gains and continued growth can, in the end, outstrip these gains and result in a greater absolute footprint, even if intensity (i.e., per capita footprint) decreases – namely, Jevon's paradox or rebound effect (York and McGee 2015; Polimeni et al. 2008; Alcott 2005).

Since the mid-1980s, the global resource demands measured by the ecological footprint have exceeded the corresponding productive capacity of the biosphere (Kitzes and Wackernagel 2009). Gains in yield and efficiency have not kept pace with increasing demand. Such ecological overshoot depletes ecological and environmental capital and leads to the accumulation of wastes in the biosphere. These changing conditions are consistent with the dynamic model projections reported by Meadows et al. (1972) as illustrated in Figure 2.2. Multiple independent assessments of the global predicament appear congruent and further underscore the need for an alternative guiding economic paradigm to address existential risk.

3.6 Human Population Growth and Finite Singularity

Continued human population growth is inarguably a major driver in the global predicament. Human population growth is one phenomenon that exhibits superlinear growth (West 2017). That is, fitting a line to the log of population size plotted against the log of time (years) produces a slope greater than 1.0. In other words, the human population is growing faster than exponential – with decreased successive intervals required to double population size (Figure 3.3). For context, as described by Cohen (1997), the world population around 2,000 years ago was approximately 250 million. The population increased to more than one billion by 1830. Only one century was required for the population to surpass two billion by 1930. The population doubled again to four billion in only 44 years. It took from the beginning

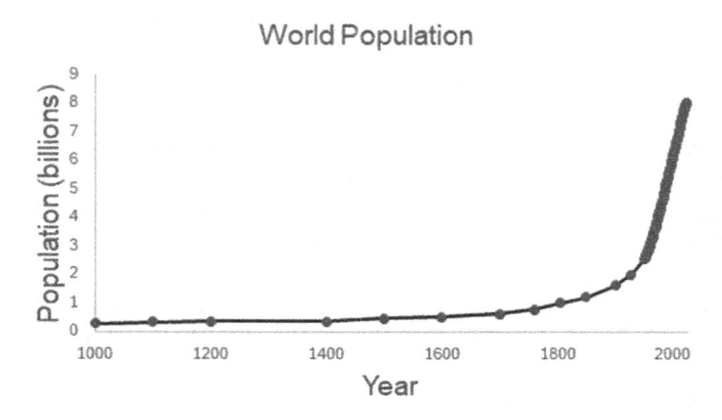

FIGURE 3.3
Superlinear human population growth.

of human history to 1830 for the population to reach one billion; adding the latest billion took 12 years (Cohen 1997).

One of the basic implications of simple exponential growth is the singularity – a point in time at which the demand for one or more critical resources to sustain growth becomes infinite (West 2017). The welcoming news for exponential growth is that the singularity occurs at infinite time. However, for growth that is faster than exponential (superlinear), the singularity occurs in finite time. Certainly, the likelihood of observing an infinite resource demand (e.g., barrels of oil, tons of grain, liters of water) seems unimaginable in real-world terms. But the finite singularity is an unavoidable mathematical consequence of superlinear growth. If global population growth continues to accelerate, something like the finite singularity and its attendant consequences for human survival might be experienced sooner than later.

Given fair warning of the mathematically unavoidable catastrophe implied by superlinear population growth, a reasonable response might perhaps take the form of some estimate of the Earth's capacity to support a human population. The question of how many people can live sustainably on the planet is not new (Cohen 1997; Malthus 1798). Numbers proposed over centuries of addressing this challenge range several orders of magnitude with higher numbers (tens of billions) tending to be justified in terms of divine providence (a deity will somehow come to the rescue). Lower numbers – lower than the current population of some 8 billion+ – deriving from some considerations of limiting resources (Cohen 1997).

Biophysical economics underscores the importance of viewing an economy as an open dissipative system with a fundamental reliance on energy inputs (Hall and Klitgaard 2018). Correspondingly, an estimate of human population carrying capacity might be developed from a basic consideration of energy required for an acceptable standard of living and total human-orchestrated planetary energy. The following describes an estimate of a sustainable global population based simply on allocation of total global energy production according to the energy needs for an asymptotic-based living standard (GJ/capita).

The United Nations Development Programme (2015), based on evaluation of the Human Development Index, suggests that an acceptable standard of living can be accommodated by about 100 per capita GJ/y (Figure 3.4). The International Energy Agency (IEA 2019) reported annual global energy production for 2019 at 617 EJ (or 6.17×10^{11} GJ) If we allocate the global energy production to 100 GJ equal increments, the global annual energy production would support 6.17 billion people – a very rough estimate of a sustainable global population assuming the ability to sustain global energy production. Of course, this crude calculation does not consider other potential basic limiting factors (e.g., food, water, shelter). Presumably, similar calculations could be made if quality of life relations can be derived for these other basic life requirements.

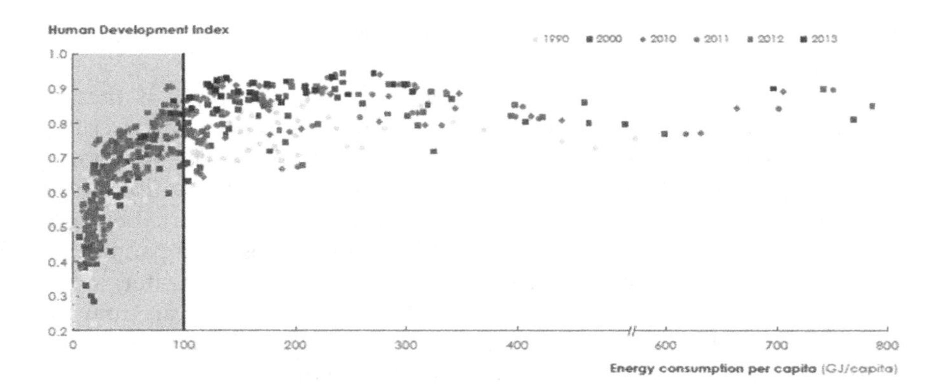

FIGURE 3.4
Standard of living (Human Development Index) as a function of per capita energy consumption (United Nations Development Programme 2015).

The takeaway points are two. One, it appears possible to arrive at some crude estimates of a sustainable global population based on quite pragmatic considerations of basic life requirements as a complement to some of the more sophisticated mathematical explorations in determining a human carrying capacity for the planet (e.g., Cohen's and others work). The rough estimate and associated uncertainty might nonetheless provide sufficient information to inform the design of a global-scale biophysical economy. Two, even the global population approximation of around 6 billion people based on inferred relationships between energy and quality of life suggests that the planet is already over-subscribed at some 8+ billion as this book heads to press. This estimate of a sustainable human population also appears to support the contentions that our planetary ecological footprint exceeds planetary resources (e.g., Rees 1992; Wackernagel and Rees 1996; Wackernagel et al. 1999). Multiple approaches could be used to estimate a sustainable global population, including global-scale simulation models (Hughes 2019, 2009).

This volume focuses initially on the design and implementation of a biophysical economy at a nation scale, namely the United States. True sustainability only has meaning at a global-scale and, in the final analysis, a biophysical economy for the entire planet will be required. The underlying assumption in this presentation is that the approach suggested for developing a biophysical economy for the United States can be applied universally – an admittedly naïve assumption, but perhaps the basic framework can be adapted to include the substantial differences in cultures, political systems, economies, and resources encountered at the planetary scale. Perhaps Cohen (1997) captures the essence of the challenge:

> "The real crux of the population question is the quality of people's lives: the ability of people to participate in what it means to be really human, to work, play and die with dignity, to have some sense that one's own life

> has meaning and is connected with other people's lives. That, to me, is the essence of the population problem." J.E. Cohen
>
> **(1997).**

Key takeaway points from this discussion continue the observations in Chapter 2 that not only is life on Earth a rare and exceptional event in the continuing evolution of the Universe, but that ever-increasing carbon life forms that inhabit this planet face the challenge of finite resources – including *H. sapiens*. Furthermore, human activities measurably impact the quality and resilience (sustainability) of life-support systems defined in this discussion as planetary boundaries. Establishment of a biophysical economy is offered as one possible answer to the pessimistic projections of the *Limits to Growth* prognostications that appear to be coming to pass unless preventative actions in the form of economic transition are undertaken at scale.

Section 2

Pillars of a Biophysical Economy

4

Biophysical Economics

4.1 Economics

Setting the stage for consideration of an alternative to mainstream economics might benefit from some, albeit brief, treatment of the subject of economics. The chapter is not offered as a course in economics. Yet the following descriptions and discussions might help to begin the conversation concerning the feasibility of implementing an economy that is guided by principles based on biophysics. Economics is first addressed in more general terms. Distinctions among mainstream economics, ecological economics, and biophysical economics are offered to provide context and further stimulate the conversation and begin to ask the key questions – is it possible to build a biophysical economy? If so, how might it be designed and constructed? If a biophysical economy was made operational, how would people know?

The word "economics" derives from the Greek *oikonomikos* and has two parts: *oikos* meaning home and *nomos* meaning management – attributed to the 4th-century Greek writer Xenophon and meaning household management (Backhouse and Medema 2009). A concise definition of the science or discipline of economics proves more difficult to pin down among practicing economists. The definition has been evolving during the past centuries of economic studies. The following discussion of the fundamental nature of economics is largely paraphrased from the extensive treatment provided by Backhouse and Medema (2009).

The scale of economics was elevated to the level of the nation in the 18th century by defining economics as political economy. Political economy expanded beyond the consideration of national wealth to address the production, distribution, and consumption of wealth. Key contributors in the development of political economy during this period include Adam Smith, Jean Baptiste Say, David Ricardo, and John Stuart Mill. Importantly, Mill considered political economy and the accumulation of wealth to derive from fundamental laws (e.g., diminishing returns, the population principle) presumed to be true (Backhouse and Medema 2009). Economic studies explored the logical deductions concerning wealth that derived from these presumed

DOI: 10.1201/9781003308416-6

laws with recognition that these laws, while assumed true, were intended mainly as guiding principles or tendencies.

The 19th century witnessed the introduction of the individual into the economic discussion. Political economy might rightly concern itself with the interactions among individuals within a larger social framework. However, economics per se became increasingly focused on human psychology and the nature of choice expressed by individuals to increase their overall welfare, where welfare was defined and measured as utility (Backhouse and Medema 2009). Key contributors to this development include Carl Menger, William Stanley Jevons, Henry Sidgwick, and Alfred Marshall. Marshall's work was central to the study of economics in the years to follow and he emphasized economics as "the study of mankind in the ordinary business of life … it is the study of wealth … and importantly the study of man," as cited in Backhouse and Medema (2009). Marshall's work also comes at a time when "economics" was becoming the increasingly favored term for this area of study, displacing "political economy."

The 1930s brought the work of Lionel Robbins, who defined economics as the science which examines human behavior in relation to desired ends amidst scarce means. This definition was not generally accepted at the time with Marshall's perspective dominating the definition and practice of economics. At the same time, increasing emphasis was placed on the empirical analysis of industry as a complex interaction between human practices and relationships – known as Institutionalism, which was offered to counterbalance more abstract theories directed at individuals maximizing their welfare (utility).

The period following World War II experienced a resurgence of Robbin's definition, which again focused more on the issue of human choices in the face of scarce resources. This period also saw the emergence of equilibrium theory and corresponding analytic activities that focused on game theory and operations research. This emphasis on equilibrium resulted in part from the collaboration of economists with scientists and engineers in addressing logistical challenges during the war. By the 1970s, the literature had moved towards the scarcity-based definition reflective of Robbins. Paul Samuelson offered a definition that economics was the study of how people of varying groups choose among scarce resources that had alternative uses in relation to costs and benefits and improved overall resource allocations (Backhouse and Medema 2009).

Based on their extensive review, Backhouse and Medema (2009) concluded that economics is variously the study of the economy, the study of the effects of scarcity, the science of making choices, and the study of human behavior. They further assess, based on their exhaustive analysis, that lack of agreement concerning a strict definition of economics perhaps does not really matter in its daily practice. These authors sympathize with a quip allegedly offered by Jacob Viner, that "economics is what economists do." For the purposes of this presentation, economics simply describes how humans conduct their daily affairs to meet material and non-material needs.

4.1.1 Historical Economies

Regardless of definitions, the design and construction of a biophysical economy might be usefully informed through some understanding of the previous schools of economic thought. The following draws selectively from the more detailed discussion of the history of economic thought provided in Chapter 2 of Hall and Klitgaard (2018) to which the interested reader is thusly referred. The purpose of the vignettes presented here is to provide (largely ecologically oriented) audiences lacking historical economic perspective with some awareness that there is a history of how society progressed through several somewhat distinct periods of economic activity and arrived at the current neoclassical economic model.

The review begins with the feudal period. As described by Hall and Klitgaard (2018), there was not much economic evolution in thought or practice during this medieval period in Europe. Wealth was determined primarily by land ownership and land was mainly owned by the nobility and the church. Religious and social conventions served to thwart social mobility and largely maintained the status quo. The feudal economy was strictly hierarchical with the nobility and the church at the top, a small group of artisans and merchants in the middle, and a majority of landless peasants, who paid taxes to the landed nobles and tithes to the church, firmly entrenched at the bottom. Economic growth was not a hallmark of the feudal period.

The thousand-year reign of feudalism began to transform in the early 16th century as merchants increasingly influenced society and the economy (Hall and Klitgaard 2018). Commerce prospered in the form and technological innovation of long-distance trade. Wealth was redefined from land ownership to the accumulation of precious metals and other material treasures primarily through trade – the age of mercantilism was in the offing. Mercantilists posited that value was defined by the price of exchange. These traders importantly intended to control the nature of exchange largely through colonial expansion, treaties, and armed conflict where necessary. Trading terms were imposed mainly against the exporter and defined to favor those who imported the goods and treasure. A positive balance of trade was the objective which was achieved largely at the expense of free trade (Hall and Klitgaard 2018). The overall goal was the accumulation of wealth (treasures) derived from trading. Mercantilists valued trade first and foremost, followed by manufacturing, and lastly, by agriculture.

The latter part of the 18th century saw a shift from mercantilism to classical political economy (Hall and Klitgaard 2018). A French school of natural philosophers (Physiocrats) emphasized a theory of value that emphasized the productivity of land as developed through labor. Nature was fundamental to value creation, particularly in relation to labor that transformed products of nature to commodities that could be sold. Contrary to mercantilists, the classical political economists espoused a laissez-faire approach as opposed to regulated trade. The political economists also based their evaluation of wealth and value in terms of production facilitated through labor, particularly the division of labor, counter to the mercantilist emphasis on value determined by exchange. Hall and Klitgaard (2018) identify key political economists as including Adam Smith, John Stuart Mill, David Ricardo, Thomas Malthus, and Karl Marx. The Physiocrats appear to have shared a concept of economics more closely aligned with that of the more recently developed biophysical economics.

Historical economics underwent a dramatic transformation with the emergence of neoclassical economics in the 1870s (Hall and Klitgaard 2018). Key contributors such as Leon Walras, Stanley Jevons, and Carl Menger maintained that self-regulating markets, perfect competition, and flexible prices should determine the detailed operations of the economy. Neoclassical economics emphasized marginal value – where the value of a commodity varies inversely with its abundance – as opposed to value determined by production or exchange. Importantly, the nature and dynamics of the neoclassical model could be readily described using the mathematics borrowed from the analysis of physical systems, mainly the calculus of systems at equilibrium (Hall and Klitgaard 2018). Neoclassical economics introduced the familiar, but impenetrable, concepts and mathematics of supply and demand, along with the inventions of rational human behavior and Pareto optimality. The neoclassical economists reframed value in the context of utility and assumed that individuals operated to maximize gains in utility resulting from exchanges in freely operating markets under circumstances of fair and open competition. Exchanges (buying and selling) continued until no individual could be made better in terms of well-being (utility) without another individual becoming worse off (Pareto optimality). Eventually, the concepts and mathematics of subjective (transactional) value and exchange were extended from individuals to entire economies. The elaboration of these ideas often in the form of mathematical physics, in addition to considerations of production, distribution, and economic growth, as well as the role of government in stimulating economic activity, contributed to the continued evolution and modern-day practice of neoclassical economics (Hall and Klitgaard 2018).

The preceding paragraphs, while brief, are offered mainly to underscore that the development and study of economies and day-to-day economic activities reflect present and past living experiences and scholarly perspectives concerning real-world historical political, economic, and social conditions. The development of ecological economics and biophysical economics, as outlined subsequently, did not originate in a conceptual economic vacuum, and these newer economic ideas importantly address intellectual and material shortcomings in economic theory and practice evident for several centuries.

This excursion into historical perspectives on economics and economies at least provides an entry point into the nature of the discourse involving historical economies and economics. It is recognized that readers of this book series who are ecologically or environmentally inclined might nonetheless be acquainted with the science of economics and the objections to neoclassical economics based on real-world ecological and environmental considerations. However, it might be less likely that the audience will have had an introduction to some of the ideas and history of the development of economic thought and practice outlined above. This book will have served a useful purpose if it stimulates professionals trained mainly in the natural and engineering sciences to delve more deeply into the realm of economic theory and practice.

The following discussions of ecological economics and biophysical economics are offered as contrasting alternatives to historical and current mainstream economics just described. The descriptions and discussions are offered in addition to help set the stage for the design and implementation of a biophysical economy and the transformation from the neoclassical model to a sustainable economy.

4.2 Ecological Economics

Ecological economics, developed during the past several decades, emphasizes an economic paradigm shift to considerations of low entropy, functioning ecological systems and dissipative structures and processes as characteristics of economics in the real world (Melgar-Melgar and Hall 2020). Ecological economics challenges neoclassical economic theory that economic value derives from consumer preferences, purchasing power, and self-maximizing transactions in perfect markets. The development of ecological (and biophysical) economics derives in part from recognizing that conventional economics is based on unrealistic representations of markets, unsupported assumptions regarding individual (rational) behavior, and associated mathematical models ill-grounded in non-representative equilibrium theory. This current, conventional paradigm dominates the world economy but largely ignores the natural sources of goods and services, as well as the energy and materials required to bring them to markets.

Melgar-Melgar and Hall (2020) provide a detailed historical description of the evolution of ecological and biophysical economics, which share similar conceptual foundations. The intent is not to reproduce their detailed account here. A principal distinction attributed to ecological economics, apart from biophysical economics, is the focus on estimating the monetary value of ecological goods and services within the current economic system (e.g., Costanza et al. 1997). In a sense, ecological economics attempts to inject ecology into the parlance and valuation of mainstream economic theory and practice. Critics of ecological economics and its valuation approach suggest that the corresponding monetary values derived for ecological resources are primarily transactional in nature and do not reflect the real-world value (e.g., costs of substitutes, embodied energy) of these goods and services. Proponents of these value estimates counter that the monetary values assigned to natural resources provide metrics that are readily understood in the discourse of conventional economics. These exchange values are contended to at least provide entry points of ecological goods and services into conversations, decision making, and policy that address the integrity and sustainability, not only of these goods and services, but as well the overall viability of economic activity as currently practiced. From another perspective, ecological economics can be viewed as a conceptual and operational approach to move the environmental – and one might argue the socioeconomic and political – consequences of the current neoclassical model from the margins towards a central focus in economics.

4.3 Biophysical Economics

Biophysical economics (BPE) builds upon the shortcomings of neoclassical economics, as underscored by the ecological economists (Odum 1971; Georgescu-Roegen 1971; Daly 1977; Daly and Farley 2011) and approaches the transformation of natural goods and services to wealth (or value) from the perspective of energy and material flows (Hall and Klitgaard 2018). Biophysical economics recognizes the duality of ecology and economics as suggested by Janus, the two-faced god celebrated by the Romans (Figure 4.1). In a sense, ecology and economics are sides of the same coin. Ecology addresses real-world constraints on economic activity in relation to available and sustainable energy and material resources; economic activity, in turn, defines a corresponding relevant ecology (Moore 2015). The practice of ecology does not occur independent of the prevailing economic context.

FIGURE 4.1
Ecology, economics, and Janus – the two-faced Roman god.

If nothing else, ecologists measure what they are able and capabilities for observation are determined in no small part by technological advances afforded by a growing and vibrant economy. More importantly, biophysical economics integrates economics and ecology with an associated expectation of sustainability (maybe eudaimonia) and reframes the neoclassical paradigm in the context of limited planetary life-support systems.

Biophysical economics is an integrated ecological and economic system that attempts to align economic growth with rates of natural resource abundance and replenishment as facilitated and constrained by available and advancing technology. The following brief discussion draws heavily from the seminal works of Hall and Klitgaard (2018) and Melgar-Melgar and Hall (2020). The beginnings of BPE trace back to at least the emphasis placed on land as a source of wealth by the 18th-century Physiocrats. This French conceptualization of economics emphasized that wealth derived principally from surplus production. The Physiocrats considered the economy to derive from combined moral and physical Natural Law, which they held as more important than free will (Quesnay 1765).

The thinking of the Physiocrats was extended in the 19th century through the discovery of laws of thermodynamics, namely that the total quantity of energy is conserved (first law), while the total quality of energy is not conserved (second law). From basic considerations of thermodynamics, evolution was perceived as driven by competition among organisms to efficiently capture energy (e.g., Lotka 1922). As stated in Melgar-Melgar and Hall (2020), Lotka (1925) introduced the term "biophysics" and developed theories (e.g., maximum power principle) that helped lay the foundation for quantitative ecosystem science and energetics. In turn, ecosystem science and energetics (thermodynamics) provide the conceptual framework for the derivation of biophysical economics.

Subsequently, the creation of wealth was considered to derive fundamentally from biophysics and was determined by thermodynamics according to Soddy (1922) – a Nobel Prize winner in chemistry. Importantly, Soddy underscored the shortcoming of conventional economics as inventing wealth that had no physical dimension with an associated disregard for thermodynamics and sources of real wealth (Melgar-Melgar and Hall 2020). One example might be the creation of money out of thin air by central banks, money which is then loaned at interest. The money needed to repay the loan and interest, however, cannot be similarly invented, but comes at the cost of work and value derived from real-world resources.

The development of BPE was also importantly influenced by contributions from systems ecologists – most notably the brothers Odum (Howard T. and Eugene). Howard Odum (1971) applied his understanding of networks and energy flows to describe the interrelations of social and natural systems. His book, *Environment, Power and Society* – and its introduction of the maximum power principle, furthering the work of Lotka – remain as cornerstones of current BPE thought. His brother Eugene in 1971 published one of the premier textbooks, *Ecology*, that remains a staple of modern ecological curricula in universities worldwide. To no surprise, notable students of H.T. Odum, including Robert Costanza, John Day, Charles Hall, and others, continued the legacy and further expanded the nature and content of BPE. An electronic search on any of these names will identify books and articles critical to the derivation and elucidation of concepts and methods underlying BPE.

For example, Costanza (1981, 1980) is recognized for showing the relationship between energy and the dollar value of goods and services in the US economy. His theory of embodied energy maintains that the value of any good or service can be derived from consideration of the energy required for their production. Costanza contributed importantly to the development of the field of ecological economics and the valuation of ecosystem services (Costanza et al. 1997) – a work that continues to generate controversy.

Charles Hall, similarly, a student of H.T. Odum and a capable systems ecologist in his own right, began his assessment of energy through studying the bioenergetics benefits and costs experienced by migrating fish (salmonids). He importantly introduced the concept and the term "energy return on energy invested," or EROI (sometimes EROEI). Hall and his students subsequently used EROI in the analysis of fossil fuels as an instrumental contribution to the continued development of BPE (Hall et al. 1981, 1979; Cleveland et al. 1984). Hall was the principal author of two books that are foundational in the development of BPE: *Energy and Resource Quality: The Ecology of the Economic Process* (Hall et al. 1986) and *Energy and the Wealth of Nations: An Introduction to Biophysical Economics* (Hall and Klitgaard 2018). This author was drawn to undertaking this volume, in part, by having the opportunity to review chapters of the second edition of *Energy and the Wealth of Nations* edition during its production.

Melgar-Melgar and Hall (2020) further recognize the contributions of Jansson (1984) in organizing a symposium in Sweden that focused on the integration of ecology and economics, again drawing on a firm foundation in biophysics. An overview of biophysical economic thought during the period 1865 to 1940 can be found in *Ecological Economics: Energy, Environment, and Society* by Martinez-Alier (1987), who emphasized the need to understand energy flows that power the economy to more meaningfully study economics. Melgar-Melgar and Hall (2020) also described several methodologies that are central to the application of biophysical economics, which are briefly outlined (Table 4.1).

The metrics and methodologies commonly associated with BPE will undoubtedly prove foundational to the design and implementation of a biophysical economy. An exhaustive discussion of each lies beyond the intention of this chapter. However, additional detail and references for each are provided (see its Table 1) in Melgar-Melgar and Hall (2020).

This brief excursion into the history and development of BPE was not intended as an exhaustive presentation, but rather as an introduction

TABLE 4.1

Metrics and Methodologies in Support of Biophysical Economics (Adapted from Melgar-Melgar and Hall 2020)

Metric	Description	Method
Energy return on investment (EROI)	Ratio that compares useful energy with energetic costs of extraction and production	Evaluates energy returned to users in relation to energy inputs required to generate energy
Social metabolism (SM)	Analysis of biophysical opportunities and constraints in metabolism among societies	Transformations and consumption of stocks and flows of energy and materials in a system
Multi-scale integrated analysis of societal and ecosystem metabolism (MuSIASEM)	Describes the biophysical feasibility and viability of a socioeconomic system	Integrates assessments of energy and material use across different scales and dimensions
Environmentally extended input-output analysis	Measure of energy and material required to provide goods and services, including environmental impacts	Uses direct and indirect inputs of energy and materials to quantify embodied energy in goods and services
Hubbert curves for depletion analysis	Measure of "boom and bust" cycles of resource production	Based on Hubbert linearization of differential equations to generate bell-shaped curve of resource production
Emergy analysis	Quantification and valuation of energy required to perform work	Measures energy used directly or indirectly to produce a product or provide a service
Ecological footprinting	Measure to quantify the biotic and abiotic resources required to sustain a human system	Uses world average productivity and global hectares to quantify spatially equivalent resource requirements

sufficient to provide context and set the stage for a discussion of how one gets from a description and justification of BPE as a necessary alternative economic paradigm to a recipe for the actual design and implementation of a working biophysical economy – the main theme of this book. To gain a greater in-depth understanding and more detailed description of BPE – its origins, development history, and current status – by all means consult Melgar-Melgar and Hall (2020) and Hall and Klitgaard (2018), as well as the myriad of references cited in these key works.

5

Creating a Biophysical Economy

5.1 Towards an Initial Biophysical Economy

The previous chapter touches on the rich history in the development of the basic description of biophysical economics and justification of its need. Less attention has been directed towards the specific design and implementation of an actual biophysical economy. Hence this book. Chapter 5 simply begins the conversation concerning the feasibility of designing and implementing an economic model that aligns with the fundamental aspects of biophysical economics.

Hall and Klitgaard (2018) extensively describe biophysical economics and justify the need for this paradigm as a replacement to the current neoclassical model as a recipe for sustainability. However, their presentation falls short in prescribing how a biophysical economy might be constructed and/or monitored. What would a biophysical economy look like in terms of specific environmental, human, political, and economic structures, institutions, and processes? How might participants know if a biophysical economy was in operation? The following outlines one possible pathway towards a US economy more closely aligned with biophysics than the current neoclassical system. The initial assumption is that the current neoclassical system can be modified to build a sustainable economy – wholesale economic reconstruction can be avoided.

One premise underlying the development of a biophysical economy is that the likelihood of success is inversely proportional to the number of current economic institutions and processes that need to change. Correspondingly, the architecture of a biophysical economy will be developed from this premise with an emphasis on making as few changes as possible. It is entirely possible that, despite this intent, wholesale reconstruction of the current economy might prove necessary to achieve a self-sustaining and flourishing biophysical economy.

Hall and Klitgaard (2018) introduce a minimalist economic model that extends the current household and firm paradigm by explicitly recognizing the open and dissipative nature of the economy which depends on inputs of

DOI: 10.1201/9781003308416-7

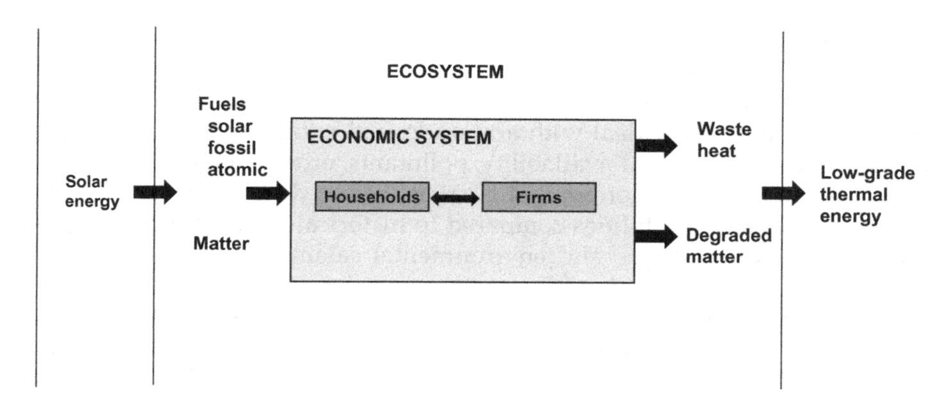

FIGURE 5.1
A minimalist approach to describing a biophysical economy (from Hall and Klitgaard 2018).

high-quality energy and matter that are subsequently transformed by the economy to waste heat and matter (Figure 5.1).

5.2 The Reality

It is important to recognize that the biophysical systems that support our current economic enterprise have been (for millions of years) and continue to operate – in a real sense we do have a biophysical-based economy. We simply do not realistically account for these economic life-support systems in our current neoclassical economic paradigm. The assumed continuous growth model has worked so far. Waves of technical advances (e.g., six Kondratieff waves) have carried the day in stimulating more or less continued growth and providing some measure of rescue from economic recession or depression (Batty 2015). These technological waves include:

- 1st wave – steam engine, ca.1780–1830
- 2nd wave – railways, steel, ca. 1830–1880
- 3rd wave – electrification, chemical industry, ca. 1880–1930
- 4th wave – automobiles, petrochemicals, ca. 1930–1970
- 5th wave – information and communications technologies, ca. 1970–2010
- 6th wave – nano- and biotechnology, artificial intelligence ca. 2010–20xx

The continuous growth model, thus far abetted by continuous technological advances, works as long as it works. The key assumption is that waves of technology will continue to support economic growth indefinitely. A counterproposal to the fundamental premise of this book would be to continue to adhere to the neoclassical model and assume that technology will save the day and effectively deal with any undesired externalities (e.g., climate change, water quality and availability, pollutants, other planetary boundaries). Indeed, Page (2005) offers that current society differs substantially in terms of technical capabilities compared to historical civilizations that succumbed to socioeconomic and environmental calamities, as described by Diamond (2005).

The following paragraphs identify the key components involved in the transition to a biophysical economy and preview the design and implementation of such an economy. Components of these topic areas will both impact and be impacted by the design and implementation of an alternative and sustainable economic system. The topic areas and their interrelationships relative to an economic transition are further articulated in subsequent chapters.

5.3 Earth Life-Support Systems

The biogeochemical processes that permit and support carbon-based life forms on Earth have been operating for millennia at planet-level scales and have only recently (i.e., Industrial Revolution) been impacted in major and measurable ways by human activities.

In response to Page (2005), the principal challenges lie in the scale and magnitude of current challenges that require a technological fix, assuming that technology will in the end suffice. For example, direct carbon capture and removal from the atmosphere continues to be developed as one technological answer to reducing the ever-increasing atmospheric CO_2 concentrations. However, to reduce atmospheric CO_2 concentration by ~1 ppm, about 2.13 gigatons of carbon must be removed from the atmosphere and removed from the global active carbon pools (source: Carbon Dioxide Information and Analysis Center, Oak Ridge National Laboratory). This equates to a cube of solid carbon of about 0.98 km on edge. The Pre-Industrial Revolution (ca 1750) atmospheric CO_2 concentration was ~280 ppm. Current Mauna Loa data indicate a CO_2 concentration of 412 ppm. The difference of 132 ppm describes aggregate modern economic activity in the age of fossil fuels. It is not clear that the equivalent of all 132 ppm need to be remedied to reverse climate change impacts, but the challenges facing any direct carbon capture and storage technologies to have a measurable impact on global CO_2 appear nonetheless daunting. A realistic question concerns whether to undertake the necessary scaling (billions of metric tons of CO_2 removal) of carbon capture and storage at a likely cost of trillions of dollars or devote some portion of those financial resources to the hardening of current infrastructure to ameliorate the projected (and currently experienced) socioeconomic and environmental (e.g., planetary boundaries) impacts of climate change.

5.4 Private Sector

Corporations and large and small businesses have the ability to act in the short term to catalyze the transition from the current neoclassical economic model to a sustainable economy compatible and consistent with operating biogeochemical life-support systems. The private sector demonstrates the capacity to modify its business activities to achieve greater efficiencies in providing goods and services and in reducing the overall life-cycle impacts of its products. Increasingly, the basic tenets of ESG (Environment, Social, Governance) are guiding management approaches towards more sustainable business models. Organizations, including the Corporate EcoForum (CEF), routinely publicize the activities undertaken by individual corporations to increase efficiencies, reduce/recycle wastes, diversify employee participation, and act in an overall responsible manner in response to concerns expressed by shareholders and stakeholders. Collaborations among private businesses, government organizations, and NGOs to operate sustainably are also featured in CEF publications as well as reported by other private sector outlets.

Corporations and businesses are able to effectively integrate their activities throughout supply chains to increase the scale and extent of sustainable business practices. The challenge remains to determine if the combined impacts of corporate actions implemented to address global change can achieve sufficient scale to make a measurable difference. Recent investigations have pinpointed the sources of GHG emissions to fewer than 100 major sources. Can the remaining thousands of global corporate entities effectively counter these sources through coordinated business activities based on sound ESG principles and practices? Will "market forces" in the spirit of Adam Smith prove sufficient to galvanize private sector actions in designing and operating a sustainable economy? Or will significant incentives (e.g., tax credits) be required from the public sector to stimulate the needed revisions to be carried out by the private sector?

Business not only impacts planetary life-support systems directly through its operations, but also indirectly exerts influence through its impacts on patterns of product consumption by its customers. Experience underscores the importance of consumer spending in driving the current economy. Importantly, the private sector variously shapes consumer spending habits through well-designed marketing and advertising campaigns. The success of any economic transition will be determined in no small part by changes in consumer preferences towards the purchase and use of commercial goods and services whose overall life-cycle characteristics are consistent with the goals and objectives that form the foundation of a biophysical, sustainable economy.

5.5 General Public

Regardless of personal value systems, people must achieve some credible level of technical understanding of how the world functions to make intelligent decisions regarding personal behavior and life choices that are compatible with a sustainable economy. The importance and responsibility of public, private, and government organizations in contributing to the education of the general populace concerning population, climate, energy, finite resources, and continuous growth economics, for example, cannot be overstated. Democracy tends to be governed according to the level of education demonstrated by a majority of the voters. When the majority educational prowess fails in the basic understanding of how the planet functions and how economies function, there is scant reason to believe that decisions necessary to achieve economic transformation to a sustainable path forward will obtain.

From a citizen perspective, the implementation of a biophysical economy, which emphasizes planning and implies limits, presents challenges that go to the very core of private property rights recognized as fundamental since the very beginnings of society and governance in the United States. Planned economies have not faired well throughout history – not necessarily because of poor planning, but more often because of greed and corruption on the part of those individuals entrusted to execute the plans. Additionally, with human cognition comes the belief that anything is possible and the concept of limits to anything runs counter to the boundless aspirations that characterize our species. In the words of Costea et al. (2007), humans have increasingly decided to "derecognize finitude." Real and measurable limits to growth cannot be meaningfully addressed when the commonly held belief is that there are no limits – and emphatically, there can never be any limits because of "boundless human resourcefulness" (Costea et al. 2007).

5.6 Government

Organizations and institutions that constitute local, state, and federal governments will play a crucial role in determining the nature and success of a transition to a sustainable economy consistent with planetary biophysical life-support systems. The ability to provide and direct financing, make and enforce laws, and adjudicate grievances within a recognized legal framework empowers governments to provide a level playing field for private and public participants who might otherwise contribute unfairly and act to their advantage in the design and operation of a biophysical economy or who might work to negate any economic transformation directed towards anything but maximizing near-term growth.

Society continues to witness the actions of governments directed towards favorable economic treatment (e.g., funding, subsidies, taxation, tax credits) of renewable energy technologies and transition to a fleet of electric vehicles. The successes to date of these activities in helping to move towards a sustainable economy have arguably been mixed. Economic growth remains a fundamental force behind these shifts in emphasis on how the economy is powered in the future.

Recent decades have seen the generation of a minimal number of laws that could be interpreted as supporting the transition to a sustainable economy. The legislative emphasis, at least at the national-scale, has been on preserving and enforcing previously enacted laws (e.g., Clean Air Act, Clean Water Act, Endangered Species Act) that have been called into question in relation to a shifting political landscape. Manifesting a sustainable economy will at least require fair and objective enforcement of existing laws. Designing and building a biophysical economy may require new legislation that remains even to be conceptualized, let alone legislated.

Importantly, the government departments and agencies collect unimaginable amounts of data that describe and quantify many dimensions of human activity, including economic performance data, sociodemographic data, multiple categories of environmental and natural resources data, agricultural and timber production data, and power generation data to name a few. The design and implementation of a biophysical economy will benefit from sophisticated management and analysis of these kinds of data. Managing towards sustainability requires understanding the complex feedback mechanisms that emerge from rich and complex socioeconomic-ecological networks that define society. Novel applications of machine learning and artificial intelligence will need to be exploited to the fullest to provide the kinds of analysis necessary to address the complexities implied by a sustainable economy.

Experience underscores the tendency for governments to be slow and inefficient in operation. However, if and when governments do act, the impacts can quickly scale to generate measurable results at local, regional, and national levels. In a hierarchical sense, governments are the "big-slow" components (Giampietro and Mayumi 2018; Allen and Starr 1982) in a complex and evolving socioeconomic and ecological system. Governments have the capacity to operate at sufficient scales to generate a measurable response to climate change and other symptoms of over-subscribed demands on planetary finite resources and assimilative capacity.

5.7 Finance

The transition from the current neoclassical paradigm to an operating biophysical economy will require financing. In turn, finance stands to be impacted by the resulting sustainable economy. Massive capital improvements in infrastructure are required to transition to a sustainable economy. Many of the improvements are being undertaken to adapt to continuing and increasing impacts of climate change (e.g., damages from increasingly frequent and intense storms, floods, droughts). Billions of dollars of damages associated with or exacerbated by climate change have been tallied in the United States during recent years. These costs are likely to continue to increase absent any commitment by the government or the private sector to address climate change at scale.

The transition to renewable energy technologies will require substantial investment, even as society continues to suffer the externalities of reliance on fossil fuels to meet the majority of power generation in the coming decades. Decarbonizing nation-scale energy technology portfolios to produce measurable impacts on reducing the costs to society from climate change will come at a staggering cost.

Of potential serious concern are the implications of a transition to an economic paradigm that is not based on growth – in fact, an economy that might emphasize limited or no growth, perhaps even degrowth. The importance of continued growth in the current financial system is paramount. Investments, including the market for equities, personal retirement incomes, and other such instruments, are based on an expectation of growth.

A highly managed and regulated economy (e.g., steady-state economy) implies limits to the accumulation of personal wealth. The imposition of such limits would seem to violate fundamental rights concerning the use and disposal of personal property and likely be challenged as an illegal taking under the US Constitution. Not to get ahead of one overall conclusion of this book, the transition to a sustainable economy based on biophysics and associated limits to growth might simply be illegal. This conjecture points back to the role of government in relation to the challenge in providing novel, yet to be conceived legislation and policy.

5.8 Policy

Policy defines a statement of intent or an organized guideline to inform decisions aimed at achieving a desired outcome. Policy can be developed and implemented by governments, businesses, other organizations, and individuals. Policy will play a critical role in the design and, particularly, in the implementation of a biophysical economy. Creating the technical means to transition to a sustainable economy is challenge enough. If society lacks the will to enact the necessary policies to apply at scale the developing sustainability technologies (e.g., renewable energy, direct carbon capture), financial resources that fuel technological advances might be better directed towards hardening the existing structure and process infrastructure to reduce risk.

The continued development of policy focused on the more traditional areas of environmental and natural resource economics will remain instrumental in effecting an economy consistent with the emphasis of a biophysical economy on the management of critical planetary life-support systems (Krautkraemer 2005).

However, challenges lie in defining and bringing into play novel policies that develop from an increased understanding of the feedback mechanisms – positive (destabilizing) and negative (stabilizing) – that characterize the diverse interactions among the many socioeconomic-ecological dimensions that define a biophysical economy as a complex and evolving system.

5.9 Feedback Mechanisms

The dynamic and adaptive socioeconomic and ecological systems and the models used to represent them will feature complex patterns of positive and negative feedback mechanisms (e.g., Sterman 2000). For example, regarding the E-dimension (environment) in ESG analysis, a first step would be to take advantage of natural resource monitoring already performed by many federal (e.g., NOAA, USFWS, USGS, USEPA, USDA) and state agencies (e.g., departments of natural resources or environmental protection). For example, fisheries and timber resources are extensively monitored. The resulting data and information can be shared not only with fishers and lumber companies, but also with lending institutions and companies that build fishing vessels, sawmills, and other equipment that services these industries. Additions to harvesting capacity could be examined within a systems context that addresses continued and future resource stocks in relation to anticipated harvests. Interest rates could be adjusted in relation to the likelihood of harvests sufficient to service loans. Closing these kinds of information feedback loops, including the likely availability and cost of energy, could help establish an economic system more compatible with the principles of biophysical economics (Hall and Klitgaard 2018).

Governments might regulate resource utilization over the longer term to avoid local extinctions or severe resource depletions, but otherwise the private sector would be empowered to make decisions that permitted shorter term economic growth, where the data indicate that growth is sustainable in the sense that future resource exploitation will not be compromised.

Permissible harvesting would be closely connected to spatial and temporal varying rates of natural resource renewal (Chapin et al. 2009), which might be influenced by evolving capabilities in resource management. Similar structures and processes could be implemented to design sustainable use of non-renewable resources, where the feedback loops might reasonably include considerations of evolving technology and resource substitution.

5.10 Models

A biophysical economy, like the current neoclassical economy, will be a complex, adapting, and evolving socioeconomic and ecological system. Mathematical models will be required to help understand the dynamic interplay among system components throughout the economic, ecological, and social dimensions of the new system (Hughes 2019, 2009). There exists a highly developed literature describing the continuing development of different genres of mathematical models and computer simulations within each of the component dimensions. However, challenges remain in integrating models across social, ecological, and economic domains at the scale of an entire national economy (but see Turner et al. 2011).

Stock and flow models, input-output models, and regional economic modeling offer opportunities to work across the many dimensions (e.g., energy, resources, labor, environmental impacts, economic performance, sustainability) of a biophysical economy (e.g., Turner et al. 2011).

The demands for data collection, data analysis, and data management to support the modeling will undoubtedly challenge current human capabilities and supporting technologies. Massive amounts of data collected, managed, and analyzed in real time will be needed to inform the implementation and evaluation of sophisticated multi-dimensional models of a biophysical economy. Artificial intelligence and machine learning might be expected to increasingly contribute to advancing the ability to meet the data requirements for the application of complex models used to help with the design and implementation of a biophysical economy.

5.11 The End Game

This chapter has introduced many of the aspects that require consideration and expansion in creating a biophysical economy. Each will be further elaborated in the chapters to come. At this point in the conversation, it is not entirely clear how these diverse aspects will be interwoven to generate an initial prescription for building a nation-scale biophysical economy. The only speculation of substance is that they must or at least will.

The initial proposition was that a biophysical economy could be designed and put in place through reasonable and realistic alterations to existing neoclassical institutions and processes. Government and other organizations continue to amass data that could be mined and better understood to guide and support the development of a biophysical economy. With economic or legislated incentives, the private sector could make use of such guidance to quickly enact corresponding changes in their business structures and operations to move towards a sustainable economy.

6

Private Sector

6.1 The Private Sector

The private sector is defined, for the purposes of this discussion, as the collection of manufacturing and service organizations (corporations), including small business and sole proprietors, that contribute formally and informally to the economy. Financial organizations are treated separately (Chapter 9), although it is importantly recognized here that the structure and function of the private sector cannot be separated from financial instruments.

The private sector entities comprise the firm in the neoclassical circular model and contribute at least half of the positive feedback loop implied by this current overarching economic paradigm. In the context of this circular model, the private sector is positioned to make substantial contributions to either maintaining the "continuous growth" economic system as it currently operates or serving as an architect in the design and implementation of a biophysical economy.

6.2 Private Sector Role in Biophysical Economics

The private sector is positioned to provide the engine that transforms the current economy to one more consistent with the principles of biophysical economics. Since at least the 1980s, sole proprietors and S corps have dominated C corporations. In 2014, there were approximately 23 million sole proprietors, 7.4 million S corps, and 1.7 million C corporations in the United States (Tax Foundation, https://taxfoundation.org). However, fewer than 5% of businesses in the United States realize more than $1 million in annual revenues; fewer than 1% achieve $10 million. At the time of writing this book, Walmart remained the largest firm among companies reporting annual revenue of more than $130 billion. The key message is that while lower in overall numbers, the C corporations provide the greatest financial (and sociopolitical) power that could be leveraged towards an economic transition towards a more sustainable, biophysical economy.

DOI: 10.1201/9781003308416-8

The private sector is positioned to contribute to an economic transition at relevant scales in relation to the urgency defined by global climate change and exceedance of several planetary boundaries with others suggesting imminent transgression (Steffen et al. 2015; Rockstrom et al. 2009). In contrast to participatory forms of government, which are characterized by anything other than alacrity, private institutions exhibit the capability to make comparatively swift and effective changes in organization and operation to achieve business goals and objectives – albeit focused mainly on profit maximization.

A key challenge in realizing a sustainable economy lies in developing a compelling business argument that stimulates the private sector, particularly the power-wielding larger corporations, to reverse course from a historical attitude that recognized the biosphere as a free good/service for resource extraction and waste processing to a perspective that underscores operating within ecological and environmental planetary boundaries.

6.3 Capitalism

The purpose here is not to provide an exhaustive discussion of the history of capitalism (e.g., Koehler et al. 2015; Sachs 1999) in relation to the needed transformation to a sustainable economy, but rather to simply reinforce the recognition that the practice of capitalism will importantly impact the likely success of such transition. "Capitalism" eludes a universal and trite definition, but the term is nearly always associated with private property ownership, capital accumulation, labor for wages, competitive markets, legally binding contracts, and agreements concerning price (Koehler et al. 2015). Capitalism is central to the development of a biophysical economy because this economic system has no equal in generating prosperity – all prosperous societies are organized around private property and markets (Rodrick 2009).

Moore (2015) characterizes capitalism as depending on the "four cheaps": cheap energy, cheap food, cheap labor, and cheap money. As these four "cheaps" are the primary means for production, it is not surprising the capitalism focuses on reducing these costs – at all costs. This view is reflected in the assertion among economists that businesses contribute to society by simply making a profit, which, in turn, supports employment, wages, taxes, purchases, and investments (Porter and Kramer 2011). From this perspective, conducting business as usual sufficiently meets social needs and benefits – the company is essentially, directly isolated from its community and social context.

Capitalism necessarily concentrates wealth in the hands of those who possess capital. Through investment, money returns money to investors at no additional expense in labor or other participation. Those who do not possess

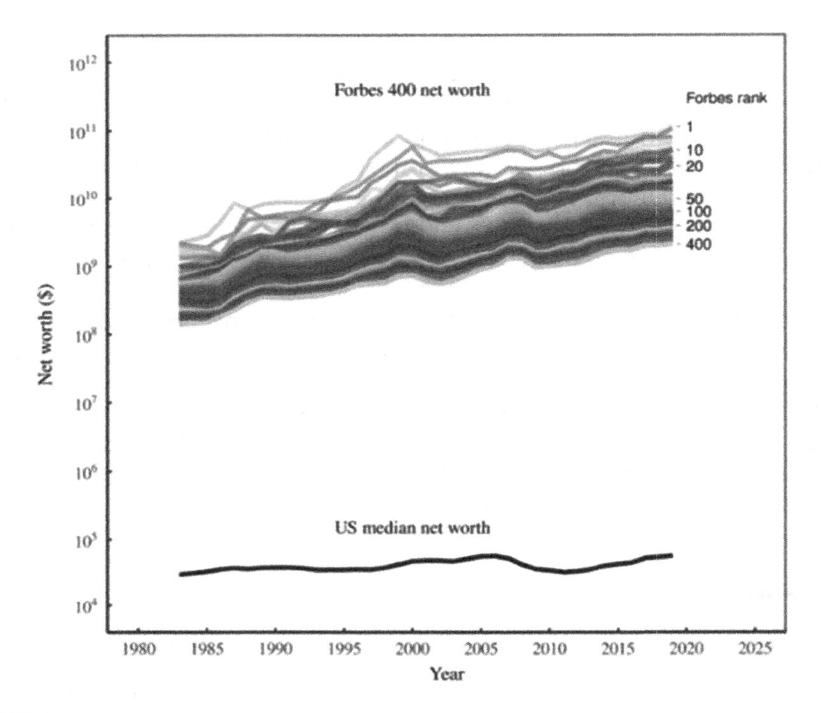

FIGURE 6.1
Net worth of Forbes 400 in context with US median net (This figure is licensed under a Creative Commons Attribution 4.0 License. Attribution goes to Blair Fix, Economics from the Top Down, https://economicsfromthetopdown.com/).

capital, through their (cheap) labor, generate the profits that wind up as "free money" in the coffers of the capitalists. Through time, the disparity in wealth can grow appreciably as demonstrated by the net worth of Forbes 400 compared to US median net worth (note the logarithmic scale in Figure 6.1).

The evident economic inequality underscores the fact that private property rights and markets require social institutions to support their functionality (Rodrick 2009). Property rights rely in legal enforcement and the courts. Markets depend on regulators to address abuse. In the words of Rodrik, capitalism is inherently "not self-creating, self-sustaining, self-regulating, or self-stabilizing."

6.4 Reinventing Capitalism

The development of a sustainable economy may require substantial reinvention of capitalism. Correspondingly, capitalism demonstrates an almost unlimited capacity to change itself (Rodrick 2009). As suggested by the economic disparities illustrated in Figure 6.1, capitalism has become increasingly perceived as a major cause of social, environmental, and economic problems (Porter and Kramer 2011). Companies continue to view value creation from a narrow perspective of optimizing short-term financial gains, while overlooking or ignoring the well-being of customers, depletion of necessary resources, supply chain viability, or economic conditions in communities that consume their products and services.

The private sector can usefully reinvent capitalism by uniting the interests of business and society (Porter and Kramer 2011). Previous and ongoing efforts concerning corporate social responsibility tend to place societal issues at the periphery of business. Porter and Kramer (2011) underscore the concept of "shared value" as transformational in business thinking. The authors define shared value as

> "policies and operating practices that enhance the competitiveness of a company while simultaneously advancing the economic and social conditions in the communities in which it operates."

Shared value addresses the combined economic and social progress through application of value principles, where value is measured in terms of benefits relative to costs, rather than costs alone (Porter and Kramer 2011). Redefining corporate purpose in terms of creating shared value, instead of simply profits, offers an approach to reshaping capitalism and its relationship to society. Such a transformation in the execution of capitalism offers a path towards increased innovation and productivity growth in the words of these authors.

Advanced democracies play an important role in the reinvention of capitalism (Iversen and Soskice 2020). Correspondingly, democratic institutions stand to play a key role in the transformation of the current economic system to one more compatible with biophysics. This transition can be increasingly stimulated through the enabling of several reforms (Iversen and Soskice 2020):

> *Financialization* that contributes to decentralized and globalized production as a hedge against economic uncertainty,
>
> *Strong competition policies* that promote innovation and allocation of resources that make difficult the establishment and entrenchment of monopolies,
>
> *Easing of capital controls and restrictions on foreign direct investments* to reinforce and expand local knowledge clusters recognized as vital to the expansion of advanced democratic economies,
>
> *Macroeconomic stability* achieved through central bank independence to promote a decentralized and globalized production structure,
>
> *Investment in higher education* where a majority of younger people obtain a university education.

6.5 Environment, Social, and Governance (ESG)

Embracing ESG as a corporate culture by executives, managers, and employees provides one pathway to increase the contribution of the private sector in reinventing capitalism and fostering the development of an economy more aligned with the principles of biophysical economics.

Business activities previously exercised under the general classification of corporate social responsibility (CSR) have further developed into corporate environmental, social, and governance (ESG) programs as briefly outlined in the following. Environmental activities in ESG commonly include assessments of corporate carbon footprints (Scopes 1, 2, and 3), energy use, water use, and waste production (Henisz et al. 2019). All companies require energy and resources to conduct business. All companies impact and are impacted by the environment.

Social components of ESG focus on the impacts and relationships of corporate operations on local people and institutions and stakeholders in local communities and stakeholders (Henisz et al. 2019). Labor relations, diversity, and inclusion are important aspects of the social components of corporate ESG programs.

The practices, processes, controls, and procedures internal to business operations define the governance attributes of corporate ESG. These activities are used to govern the firm itself, as well as to comply with regulations and address the needs and expectations of external stakeholders (Henisz et al. 2019).

Multiple independently constructed ESG frameworks (e.g., Sustainability Accounting Standards Board, Global Reporting Initiative, Carbon Disclosure Project, International Integrated Reporting Council) have been converging towards one internationally harmonized approach to ESG analysis and reporting. Correspondingly, alternative specifications of ES have been published (Costanza et al. 2017). However, congruence among the broader ES classifications and individual services evident across the alternatives indicate a corresponding harmonization. Integration of ES into ESG analysis and reporting should continue with the objectives of (1) delineating specific metrics used to define ESG targets across the SDGs and (2) evaluating performance across the SDGs to support consistent ESG analysis and reporting.

A growing and evolving body of technical literature continues to characterize empirical relationships between corporate ESG factors and corporate sustainability (Oprean-Stan et al. 2020; Henisz et al. 2019). Firms enact ESG management programs with the belief that competent ESG activities contribute towards corporate sustainability (e.g., Orlitzky et al. 2003; Margolis et al. 2009; Eccles et al. 2014; Kotsantonis et al. 2015). Sassen et al. (2016) reviewed the literature and concluded that firms with substantial programs in corporate social responsibility (CSR), measured by ESG factors, generally demonstrated negative relationships between CSR performance and firm risk, where risk is defined as the cost of capital.

At the aggregate level, there appears a general relationship between corporate performance and ESG programs. For example, Kumar et al. (2016) analyzed data (2014–2015) from 966 corporations across 12 industry sectors and found that firms with high scores for ESG activities consistently demonstrated decreased volatility in stock price compared with firms that performed less convincingly in ESG.

The results are less clear when examining the contributions of individual dimensions of ESG to corporate sustainability (Sassen et al. 2016). Positive relationships between corporations and local communities (S), as well as relationships with employees (G), appear negatively correlated with firm risk (Bouslah et al. 2013; Oikonomou et al. 2012). However, mixed results were obtained in analyzing corporate environmental (E) activities in relation to corporate performance (Bouslah et al. 2013). Similarly, Kim and Li (2021) statistically analyzed corporate MSCI ESG data and S&P Capital IQ-Compustat financial performance data for 4,708 firms representing a wide range of industries over the period of 1991–2013. Their results based on simple correlations and multivariate regression indicated a positive relationship between ESG factors on corporate profitability with corporate governance having the greatest impact. The social factors of ESG demonstrated the strongest relationship with firm credit rating. Curiously, the environmental component of the total ESG score indicated a negative relationship with credit rating. In contrast, Park and Jang (2021) found that environmental and governance factors were more important in influencing investment decisions in an analysis of South Korean corporate performance and ESG data.

In addition to empirical analysis, dynamic simulation models can be used to explore causal relationships between ESG and corporate sustainability (Bakshi and Fiksel 2003; Cosenz and Bivona 2021; Cosenz et al. 2020, 2019; Lozano 2015; Sterman 2000). The models attempt to mechanistically describe the structure and function of corporate systems within a broader business, social, and natural environment. The models capture the fundamental business operations of the corporation, while including specific inputs and outputs that directly connect the performance of the modeled firm to dimensions of ESG.

6.6 Sustainable Development Goals

Firms have become increasingly invested in using United Nations sustainable development goals (SDGs) to guide their corporate ESG programs. The United Nations has developed 17 categories of sustainable development goals (SDGs) offered to help guide international efforts directed towards achieving sustainability (Table 6.1.).

TABLE 6.1

United Nation Sustainable Development Goals (SDGs) and Linkages to Ecosystem, Services (ES). https://www.un.org/sustainable development

SDG	Example SDG Targets, by 2030...	Relevant ES[a]
1-No poverty	eradicate extreme poverty; implement social protection systems; ensure access to resources and basic services	provide biochemicals and natural medicines
2-Zero hunger	end malnutrition; ensure sustainable food production systems; double agricultural production; maintain genetic diversity of seeds, plants, farmed animals	provision of food, water, and fiber; genetic resources; climate regulation; water regulation; erosion regulation; soil formation; pollination; regulation of pests; nutrient cycling; primary production
3-Good health, well-being	end epidemics; reduce deaths from communicable disease; reduce deaths from toxic chemicals; research for medicines and vaccines	provide biochemicals and natural medicines; air quality regulation; natural hazard regulation; water purification; waste treatment; regulation of pests and disease; recreation
4-Quality education	achieve youth and adult literacy and numeracy; ensure completion of primary and secondary education; increase tertiary education; eliminate gender disparities	cultural services, including aesthetics, diversity, spiritual values; knowledge systems; educational values
5-Gender equality	end discrimination and violence against women; ensure participation in leadership; women access to economic resources; empower women	cultural diversity; knowledge systems; educational values
6-Clean water, sanitation	access to drinking water and sanitation; improve water quality; implement water resources management; protect water-related ecosystems	water regulation; water purification and waste treatment; erosion regulation
7-Affordable, clean energy	access to affordable, reliable energy; increase renewable energy; invest in clean energy infrastructure	climate regulation; water regulation; primary production
8-Decent work, economic growth	sustain per capita economic growth; technological upgrading and innovation; improve efficiency of resource consumption; eradicate forced labor, slavery, and trafficking; promote sustainable tourism	cultural diversity; knowledge systems; educational values; eco-tourism
9-Industry, infrastructure	promote sustainable industrialization; increase resource use efficiency; enhance research; develop resilient infrastructure; Internet access	air quality and climate regulation; natural hazard regulation; water regulation; water purification and waste treatment; knowledge systems

(Continued)

TABLE 6.1 (CONTINUED)

United Nation Sustainable Development Goals (SDGs) and Linkages to Ecosystem, Services (ES). https://www.un.org/sustainable development

SDG	Example SDG Targets, by 2030…	Relevant ES[a]
10-Reduced inequalities	promote social, economic, political inclusion; greater equality in wages, fiscal policy, and social protection; well-managed migration policy	cultural diversity; religious values; knowledge systems; educational values
11-Sustainable cities, communities	access to safe, affordable housing and transportation systems; reduce impacts of disasters; reduce impacts of air quality and waste; accessible green spaces; regional and national development planning	air quality and climate regulation; water and natural hazard regulation; water purification and waste treatment; erosion control; regulation of pests and diseases; biodiversity; recreation; cultural diversity; knowledge systems
12-Responsible consumption, production	efficient use of natural resources; sound management of chemicals and wastes; reduce waste generation; adopt sustainable practices and reporting	water purification and waste treatment; knowledge systems
13-Climate action	strengthen resilience and adaptive capacity; integrate climate change measures into strategies and planning; improve education and awareness for mitigation, adaptation, and early warning	climate regulation; natural hazard regulation; water regulation; knowledge systems
14-Life below water	manage, conserve, and protect coastal and marine ecosystems; reduce marine pollution; minimize ocean acidification; regulate harvesting; restore fish stocks to produce maximum sustainable yields	climate regulation; nutrient cycling, photosynthesis, primary production; biodiversity; food provisioning; knowledge systems
15-Life on land	conserve, restore, and sustain terrestrial and freshwater ecosystems; halt deforestation; increase afforestation; halt loss of biodiversity; share benefits of genetic resources; control invasive species	climate regulation; nutrient cycling, photosynthesis, primary production; water regulation; soil formation; erosion control; biodiversity; pollination; regulation of pests; genetic resources; knowledge systems
16-Peace, justice	reduce all forms of violent death; end exploitation, trafficking, and torture; reduce corruption and bribery; develop accountable and transparent institutions	cultural diversity; spiritual and religious values; knowledge systems; educational values
17-Partnerships	Mobilize financial and technical resources for developing countries; promote equitable multilateral trading; increase exports to developing countries; enhance policy coherence for sustainable development	knowledge systems; educational values

[a]*Based on ES categories in MEA (2005)*

FIGURE 6.2
Example mapping of individual SDGs to the three dimensions of corporate ESG programs.

The SDGs can be mapped onto the three individual dimensions of corporate ESG programs and used to define goals, objectives, and measurable targets for each dimension of corporate ESG programs (Figure 6.2). Sustainability might be characterized by the intersection of these ESG diagrams.

6.7 A Science of Corporations

The obvious importance of private corporations in the functioning of an economy has stimulated interest in whether corporate performance can be addressed conceptually and practically assuming a science of corporations (West 2017, chapter 9). The empirical relations between individual dimensions of ESG and different measures of corporate performance, along with the example application of causal models for corporate sustainability assessment, imply the existence of a "science" of corporations. West (2017) analyzed corporate data compiled by Standard & Poors for 28,853 publicly traded firms over the period of 1950–2009. This study established simple statistical relationships (log:log regressions) between firm size (measured number of employees) and corporate net income and total assets. The regression slopes were <1, which demonstrates sublinear (less the 1:1) scaling, similar to how individual organisms scale with size (mass) (West 2017). The sublinear scaling provides for efficiencies of scale; that is, doubling the size of the organization can be supported by less than twice the business infrastructure. The analysis of the corporate data also revealed age-dependent growth rates and non-zero mortality rates for firms of varying size and age. Analogous to organisms, corporations exhibit an initial period of rapid growth, followed by longer term slower growth (or no growth) as increasing maintenance costs accumulate with firm size and age (West 2017).

Further analysis of the Standard & Poors data provided insights concerning corporate mortality (Daepp et al. 2015). Symptoms of corporate mortality include zero annual sales, exhaustion of capital, and net present value decreasing to zero. The authors discovered that the cause of nearly half of the corporate "deaths" took the form of merger or acquisition. Only about 8% of the deaths were listed as bankruptcy or liquidation. The analysis suggested that any firm, regardless of size, age, or economic sector, has an annual ~10% chance of mortality. The data indicated some degree of liability in relation to corporate senescence or obsolescence (Daepp et al. 2015).

Pursuing ES and ESG with an emphasis on a "science of corporations" offers several opportunities. Standardized and consensus metrics can be defined to quantify individual ecosystem services selected to inform SDGs selected for the individual dimensions of specific corporate ESG analysis and reporting. Well-defined metrics can serve as the outcomes to be monitored in relation to ESG management and corporate ESG performance objectives. The metrics can importantly guide the design and implementation of rigorous monitoring programs and ensure the collection of data that can seamlessly inform ESG management and decision making. Empirical or causal (modeled) relationships between management actions and outcomes (i.e., management–response functions) permit the derivation and implementation of specific actions with expected results that relate directly to corporate ESG goals and objectives. Precise metrics, management–response functions, and robust monitoring schemes constitute the key components of adaptive management. Correspondingly, corporate ESG management, analysis, and reporting can be organized and conducted with an adaptive management (AM) framework. An AM framework is a powerful methodology for generating feedback between management actions, expected outcomes, and measured outcomes. This feedback can inform subsequent decision making aimed at achieving corporate ESG goals and objectives within a continuing and evolving management and decision-making process. The impacts of imprecise data and uncertain manage-response functions on decision making and corporate ESG performance can be explicitly addressed within the AM framework.

Ecosystem services can inform consistent ESG analysis and reporting by providing uniform metrics for target setting and performance evaluations across SDGs in relation to all three dimensions of ESG. Table 6.1 demonstrates an initial cross-walk among individual ecosystem services and sustainable development goals. This kind of mapping can help guide the selection of ecosystem services appropriate for individual SDGs and help identify corresponding data and information needed to quantify the selected ecosystem services in support of ESG analysis and reporting.

Building ESG programs underpinned by ES emphasizes the inculcation of quantitative ecology and environmental systems science in support of corporate ESG analysis and reporting. Formulating ESG goals and objectives with the explicit consideration of ES-driven metrics takes additional advantage of the continued evolution, sophistication, and coalescence of concepts and methodologies that define ecosystem services.

In the end, well-intentioned effort in the integration of ES and SDGs in service of corporate ESG programs does not necessarily confer firm sustainability,[1] if ESG activities occur within an economic context that requires continuous growth. Growth will overwhelm efficiency gains and potentially generate a larger environmental footprint than prior to the increases in operating efficiencies. Ecological niche theory reminds us that successfully addressing the most pressing limiting factor (e.g., energy) inevitably brings the next limiting factor (e.g., water) into importance in determining economic activity. The theory also emphasizes that the number of potentially limiting niche dimensions is not known *a priori*. Limitations deriving from biophysical life support systems are not particularly negotiable. Human prosperity, including corporate sustainability, on a finite planet is ultimately subject to biophysical constraints not directly addressed by conventional economics. These overarching constraints provide minimal opportunity for humans to measurably intervene at sufficient scale. The wicked challenge is to make use of the best available quantitative science of the environment, integrated with an emerging science of corporations, to derive and operationalize sustainable growth – with all its attendant benefits – on a finite planet.

6.8 The Private Sector, Sustainable Economy, and Scale

Assume that the private sector wholly embraces and manifests corporate conduct commensurate with the biophysical requirements to develop a sustainable economy. The key question then becomes one of scale. As suggested previously, the private sector evidences an ability to act in efficiently and effectively in the near term. But the question remains, for example, if the entire complex of corporate entities in the United States reduced overnight its GHG emissions to zero, would it make a measurable difference in subsequent measures of atmospheric CO_2 concentrations? Watts (2024) suggests an initial answer of yes and reported that only 57 companies are responsible for 80% of GHG emissions since 2016. These companies clearly have the ability to reduce GHG emissions at scales relevant to global atmospheric CO_2. Regrettably, Watts (2024) further reported that many of the fossil fuel–producing companies increased production in the wake of the Paris agreements.

Agricultural production in the United States in 2022 was dominated by several major companies: Archer-Daniels-Midland (ADM), Bunge Ltd., Cargill, Tyson Foods, Inc., and Louis Dreyfus. ADM reported an annual revenue of $32.6 billion (USD). In the face of increasing pressure from climate change on production, modern agriculture is relying more on automated technologies, including sensors, analytics, robotics, and automated systems (e.g., weeding robots) to maintain and increase yields and make optimal/maximal use of available resources.

Similarly, beef production in the United States is largely controlled by four companies (Reuters 2021). Negative impacts to the environment as a result of beef production and processing could be largely addressed by these companies. Steps taken by these companies to adopt procedures consistent with sound environmental practices could set the bar for sustainable beef production throughout the industry. These companies might also be able to exert significant influence in reshaping dietary preferences to reduce meat consumption in the United States and reap the environmental benefits of a human diet that emphasized plants and plant by-product consumption.

As of 2021, the top ten softwood lumber producers in the United States had the capacity to produce 24 billion board feet, which represents about 53% of the total lumber industry in the United States (Forisk, https://forisk.com). The top ten producers represent more than 1,900 forest industry mills tracked by Forisk. Sustainable forest industry procedures adopted by these ten companies could impact the forest industry at sufficient scale to shift the industry towards a net positive environmental footprint. The forest product industry offers an opportunity to manage carbon at scales sufficient to generate measurable impact on carbon dynamics in the United States.

The above are but a few examples of where consolidation of major production across key industries could provide opportunities for the private sector to make meaningful and measurable impacts on the transformation of economic activities towards a more sustainable model consistent with the fundamental principles of a biophysical economy.

Note

1. Corporations define a sustainable growth rate (SGR) as the rate of growth that can be funded internally, without borrowing (Platt et al. 1995; Higgins 1977). This purely financial definition of sustainable growth appears distantly related, at best, to sustainable growth as addressed by corporate ESG activities.

7

General Public

7.1 The Neoclassical Consumer

In the end, the biophysical economy is intended to help human beings persist and genuinely flourish – that is, achieve eudaimonia.[1] The public sector for purposes of this presentation refers to people (economic consumers) in general or the household component of the neoclassical model. The public sector can significantly influence the design and implementation of a biophysical economy through its patterns of consumption of the goods and services produced by the economy. It is common knowledge that consumer spending is a major force underlying a growing economy. In the end, the nature of the economy primarily reflects the values of the consuming public and their willingness and ability to purchase.

Fundamental to the neoclassical economic model is the concept of the household that comprises a homogeneous group of individuals that strive in common to achieve maximum well-being through rational behavior and choices (Isaac 1998). Microeconomics addresses the structure and implications of choices. At some point, the design and implementation of a biophysical economy will confront neoclassical consumer theory (NCT), abstract choice theory (ACT), and their applications. Earth system scientists, systems ecologists, and biophysical economists must necessarily become familiar with the concepts and language particular to neoclassical consumer theory and abstract choice theory to facilitate meaningful conversations about biophysical economics with participating neoclassical microeconomists (e.g., Table 7.1).

NCT views individual consumers strictly as purchasers of consumer goods. The purchase is a choice among individual bundles of goods based on consumer preference. The consumer is considered not to buy individual goods (as they actually do), but rather bundles of different quantities of individual goods.

Consumers are also assumed to demonstrate indifference to sets of bundles, where there is no preference among these bundles. If there are only two different goods in the bundles, indifference sets can be plotted as a continuous, convex function two-dimensional curve. Mathematical extensions

DOI: 10.1201/9781003308416-9

TABLE 7.1

Terminology Important in Conversations with Neoclassical Consumer Theorists

Term	Description
bundle	A bounded collection of different goods potentially selected by the consumer, (e.g., one apple, two pears, three oranges)
preference	Feelings that influence actions, namely the selection from among different bundles of goods
indifference sets	Sets of bundles among which the consumer has no preference
indifference curves	Graphed continuous functions of indifference sets
marginal rate of substitution (MRS)	Slope of a line tangent to an indifference curve at a point defined by a specific bundle
affordable set	The set of bundles that can be purchased constrained by the consumer's budget
consumer demand curve	Locus of points describing units of goods purchased in relation to price
utility function	Mathematical representation of consumer preferences
marginal utility	Partial derivative of a utility function
econometrics	Application of statistical hypothesis testing to economic models

of these indifference curves apply to bundles with more than two kinds of goods. The indifference functions permit the quantification of trade-offs among bundles defined as the slope of a line tangent to the function at the point location defined by a bundle of interest. This trade-off is defined as the marginal rate of substitution (MRS).

The consumers choice process is static. The consumer is envisioned to enter a retail outlet and decides on a purchase based on the amount of money available to spend and the prices of the goods for sale. The retail prices are defined as exogenous parameters taken as given by the consumer. In NCT, the consumer purchases affordable bundles defined as the budget or budget constraint. The selected bundle of goods derives from the preferences for bundles expressed by the consumer.

NCT uses information concerning consumer preferences and store prices to develop demand curves. The demand curve is defined by the locus of points that quantify the number of units of a particular good purchased as a function of its price. Demand curves can be constructed for all relevant goods in the market. The resulting aggregate demand curve can be used with corresponding market supply curves to define an equilibrium price – where the quantity demanded equals the quantity supplied. Thus, beginning with preferences, NCT can derive the quantity and price for an equilibrium economy, where individuals are optimizing their exchanges with others in society.

One of the stated advantages of NCT is the translation of the consumer choice problem just described to one that can be addressed through mathematical optimization. This translation is facilitated using the concept of utility. A utility function is a mathematical description of consumer preferences. The contents of each bundle define a corresponding numerical value of utility based on consumer preferences. The utility value is unitless. The utility function compares the utility of its component bundles. The consumer's optimal bundle can be defined as a utility maximization problem constrained by the consumer's budget.

The partial derivatives of the utility function define marginal utilities. These derivatives are purely mathematical devices. They have no real meaning, other than as partial derivative of a function. However, some economists interpret marginal utilities, for example, as the increase in satisfaction associated with the consumption of a little bit more of a good. Such interpretations are of dubious quality because the utility functions again compare bundles, not individual consumer goods.

The advantages of NCT are claimed to be the description of complex consumer dynamics in a structured manner, the resulting ability to translate such structure into mathematical representation using differential calculus, and the ability to characterize and analyze the consumer choice problem using the power of optimization. If differential calculus provides power and rigor in the quantification and understanding of the physics of moving objects, then the demonstrated application of such mathematics in the realm of microeconomics must correspondingly validate neoclassical consumer theory and provide a robust description of consumer choice. Economists, like ecologists, exhibit some degree of physics envy – even though physicists are blessed with challenges posed by vastly simpler mechanical systems.

NCT has been criticized. A fundamental concern is that it begins with a definition of preference developed in the context of human psychology rather than actual observed behavior. Perceived preferences can be altered in action (i.e., actual consumption) when other factors come into play, for example, peer pressure to purchase an otherwise non-preferred good. The concept of indifference and its measure in terms of utility lead logically to inaction when the consumer is confronted by bundles of equal utility. Inaction does not affect economics. Framing the choice as a selection among bundles sets up a problem that seems to have little relevance to how people actually purchase goods. Consumers purchase quantities of individual goods not bundles. However, the concept and definition of bundles of goods provide the ability to construction continuous functions for subsequent analysis using the power of differential calculus and optimization. An analogous NCT based on consumer purchase of individual goods becomes quickly intractable.

The biophysical economist is challenged to meld real-world phenomena that influence consumer choice (e.g., a finite resource base, pollution implications of certain purchases) into the concepts and mathematics of NCT or offer an alternative theory that meets the combined needs of neoclassical microeconomists and earth systems scientists devoted to achieving a sustainable economy.

7.2 Maslow's Hierarchy

Contrary to the common rational man implied by traditional economic analysis are the complexities in thought and behavior that inform the actions and choices of actual human beings in the market place. Human psychology, apart from the "physics" of neoclassical consumer theory, will importantly influence the nature and success in designing and implementing a biophysical-based economy. A comprehensive treatment of human psychology is well beyond the intent of this book. However, one example of the characterization of complex human behavior was provided by Maslow (Poston 2009; Maslow 1943). Maslow suggested a pattern and hierarchy in the motivation and manifestation of human behavior (Figure 7.1). Each component in Maslow's hierarchy appears relevant in the discussion of the economic transition implied by biophysical economics.

7.2.1 Physiological

At the most basic level, humans require food, water, oxygen, adequate rest, and a diverse array of macro- and micro-nutrients, in addition to shelter

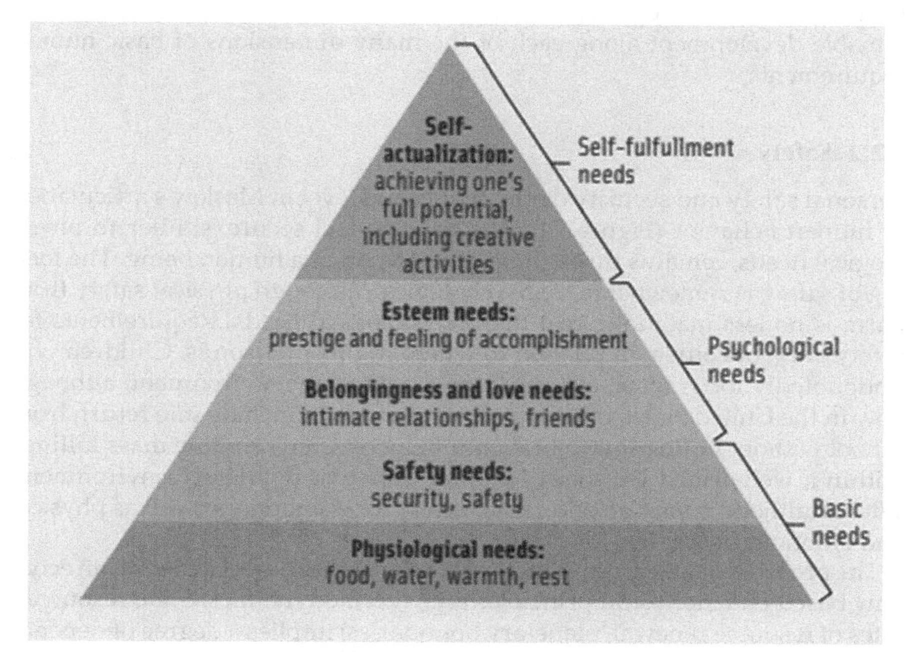

FIGURE 7.1
Illustration of Maslow's hierarchy of needs (from Poston 2009).

from the elements, to sustain a complex, homeostatic physiology (Poston 2009). Motivations to secure these basic resource needs are unsurprisingly very strong. This level of basic needs also includes the hormonal regulation of key aspects of human behavior and physiology, including the "flight or fight" response stimulated by adrenalin and reproductive cycles determined by estrogen and testosterone. Important to recognize is the fact that these needs persist, by definition, throughout the human lifespan. The overwhelming desire to meet the physiological requirements to sustain life itself is undoubtedly one of the most important motivations of human behavior and meeting these physiological needs is not surprisingly the foundation in Maslow's hierarchy (Figure 7.1).

Certainly, one of the main objectives of a biophysical economy is to ensure the long-term availability of food, water, and an otherwise inhabitable (e.g., clean air, clean water, survivable temperature and humidity) planet for humans and other organisms. Human technologies including modern industrial agriculture, water purification, buildings, transportation, heating and cooling, electricity, information processing – essentially all aspects of the modern economy – have become increasingly important in securing reliable basic resources. Accordingly, meeting the basic needs as defined by Maslow for an expanding human population will require a comprehensive transformation of the current economic model to one that addresses sustainable development along each of the many dimensions of basic human requirements.

7.2.2 Safety

Personal safety and security define the second level in Maslow's articulation of human behavior (Figure 7.1). The need to feel secure, similar to physiological needs, remains throughout the lifespan of a human being. The feeling of safety is somewhat more psychological, although physical safety from harm is no less materially real than physiological needs. Requirements for safety vary, of course, in relation to where in life a person is. Children will undoubtedly focus on a safe and functional family environment, although now in the United States this focus has expanded to include safe return from school without falling innocent victim to increasingly routine mass killings within a well-armed US society and an impotent regulatory environment. With adulthood comes an additional desire for economic, as well as physical and emotional, security.

The promise, or at least emphasis, of sustainability associated with an economy based on biophysical principles (e.g., thermodynamics, finite resources, rates of resource renewal, planetary boundaries) implies a degree of dependable resource availability to meet basic needs and an increasingly predictable economy, both of which might contribute to a corresponding sense of security among participants.

7.2.3 Belongingness and Love

The third component of the Maslow hierarchy defines a need to belong on a social level (Figure 7.1). Maslow postulates that the more social aspects of the hierarchy manifest only after the basic physiological and safety requirements have been satisfied. The human sense of belonging, as described by Poston (2009), emerges as an increased focus on establishing relationships with other people. Similar to safety and security, the exact forms of built and maintained relationships vary and evolve with life stage. Children and adolescents might value inclusion in groups of their selected peers in addition to developing a sense of belonging within their immediate family. Adults might increasingly value additional friendship, including romantic partners. Maslow suggests that building a strong sense of belonging is an important requisite for self-esteem.

A biophysical economy can directly contribute to developing a sense of belonging through the creation of novel peer groups that share common beliefs in the need and value of an economic paradigm that is consistent with the fundamental nature of planetary life support systems. There will be new groups to join. Indirectly, the economic transition to a sustainable model can contribute to opportunities for belonging by helping to meet the first two levels – physiology and safety – of the hierarchy in a sustainable and predictable manner.

7.2.4 Esteem

Self-esteem requires constant maintenance throughout life just as the previous three levels of the hierarchy (Figure 7.1). Maslow outlines two forms of self-esteem (Poston 2009). An individual's ego defines a lower form of self-esteem, which is realized as a function of established status, reputation, fame, and appreciation. This lower form requires reinforcement and validation to be maintained. A higher form of self-esteem emerges as self-respect. Once a person has gained respect for himself or herself, it is more difficult to lose that aspect of self-esteem. The higher form (essentially liking oneself) appears to require less maintenance.

Similar to contributing to a sense of belonging, a biophysical economy can help to provide a consistent resource base (at some level of productivity) that reliably serves human needs defined by the lower levels in the hierarchy. Sustaining these lower-level needs provides the opportunity for individuals to focus on the aspects of life that contribute to building self-esteem, whether it is merely maintaining ego or developing self-respect.

7.2.5 Self-Actualization

The highest level in Maslow's hierarchy identifies self-actualization (Poston 2009). Having gained self-esteem, self-actualizers focus on what matters most in defining who they are. In a sense, self-actualization is the internal conversation that everyone establishes as some point of their life. With all the previous needs met, an individual can search for their true calling in life. Self-actualization manifests more as the moral, ethical, and spiritual aspects of human nature – essentially finding one's place in the Universe.

Similar to supporting and enabling self-esteem, a biophysical economy can provide consistent and reliable sustenance at the lower levels of the Maslow hierarchy to facilitate self-actualization. Importantly, individuals who have developed self-esteem and who continually invest in furthering their own moral, ethical, and spiritual compass might well be the individuals who contribute substantially to meeting the challenges posed by the major economic transition required in moving from current conventional neoclassical economics towards a biophysical economy. These individuals are more likely to be in a position in society to speak truth to power – a key requirement in undertaking any major shift away from the current economic paradigm.

Maslow's (1943) hierarchy is not without its criticism (e.g., King-Hill 2015). Clearly, the five levels are not mutually exclusive and individuals are likely to be simultaneously active across multiple levels. Additionally, as outlined by King-Hill (2015), the hierarchy does not take into consideration other societal factors that might influence human behavior, including, for example, war and recession (Cianci and Gambrel 2003). The relative importance of the different levels in the hierarchy also appears related to individual age (Tay and Diener 2011). The main message is not to embrace Maslow's theory as the alternative to the neoclassical representation of human consumer behavior. The point is that there are alternative models of human behavior that might not lend to convenient mathematical description, but more realistically characterize choice in the context of more traditional microeconomics.

7.3 A Biophysical Consumerism

7.3.1 Spending

Consumer spending in large part drives the current neoclassical economy. Individuals who alter their preferences towards goods and services that are less demanding on the environment or who tailor their desires for resources in relation to the natural or managed rates of replenishment could measurably contribute to the implementation of a biophysical economy. Similar to the labeling of contents in consumer products, readily identifiable information concerning the environmental impacts of a product within an overall life cycle analysis could guide consumers to purchases of goods and services with lesser impacts (Kissinger et al. 2013; Huijbregts et al. 2008). Table 7.2 lists the amount of greenhouse gas emissions and magnitude of ecological footprints

TABLE 7.2

Ecological Footprints and Greenhouse Gas Emissions of Materials (from Kissinger et al. 2013)

	Average GHG Emission		Average Ecological Footprint
Category	**CO_2eq/tonne**	**Category**	**gha/tonne**
Cardboard	890	Glass	0.24
Glass	990	HDPE – plastics	0.20
HDPE – plastics	1,015	PVC – plastics	0.41
Newsprint	1,120	PET – plastics	0.48
Printing paper	1,290	Steel	0.61
PVC – plastics	1,920	Polystyrene	0.66
PET – plastics	2,240	Diapers	1.14
Steel	2,530	Newsprint	2.23
PS Tab plastics	2,970	Cardboard	2.76
Diapers	3,580	Printing paper	2.82
Aluminum	10,840	Aluminum	2.42
Cotton fabric	21,500	Cotton fabric	10.20

associated with selected categories of goods determined by Kissinger et al. (2013). Although not specified at the level of individual retail products, the information summarized in Table 7.2 can nonetheless provide some general guidance concerning the ecological and environmental implications of certain categories of manufactured products. Not surprising, aluminum products imply a substantial environmental cost largely associated with production. Hence, the high value of aluminum recycling. Somewhat startling was the highest emission and footprint for cotton fabric. These kinds of summaries and analyses (see also ecoinvent Centre 2004; Frischknecht et al. 2005) could be extended to the necessary environmental impact information to include on individual product labels for specific consumer products. Brand competition might develop along this dimension of product environmental quality and impacts.

7.3.2 Consumer Ecoregions

A basic element of business analytics lies in the quantification and geographic delineation of consumer demand (Holman and Hacherl 2021). State and county boundaries or US Postal Service ZIP codes often define geographic aggregations. However, in rethinking the role of consumerism in the design and implementation of a biophysical economy, alternative aggregations that reflect regional environmental conditions as they influence demand might prove more useful that traditional political boundaries.

Consumer ecoregions were developed by aggregating areas similar in climate, landscapes, economics, and consumer culture (Holman and Hacherl

2021; Holman 2020). The characteristics used in defining ecoregions also included similar population sizes and economic activity (i.e., regional GDP). The authors used regional temperature, precipitation, topography, extent of river networks, and solar radiation in their aggregations. Local areas gross domestic product and population were also used in defining the consumer ecoregions. The analysis was performed at the county-level of spatial resolution. Table 7.3 lists the consumer ecoregions defined for the eastern conterminous United States (Holman and Hacherl 2021). Ecoregions were similarly developed for the western United States (Holman 2020).

The intent of the consumer ecoregions is to provide insights that can inform location-based decision making for businesses and increase the understanding of the importance of boundaries other than political in shaping consumer behavior. This kind of approach that describes consumers demand in relation to regional environmental characteristics might prove useful in forecasting or shaping demand in relation to resource availability consistent with a biophysical economy.

7.3.3 Savings and Investments

In addition to spending habits, the nature and magnitude of savings and investments by the public can also impact the likelihood of setting up and maintaining a sustainable economy grounded firmly in biophysics. Financial resources directed to savings and investments correspondingly reduce near-term consumer spending and the concomitant demand for production of goods and services. Investments can be directed to providers of goods and

TABLE 7.3

Consumer Ecoregions of the Eastern Conterminous nited States (from Holman and Hacherl 2021)

Ecoregion	Counties	Population (2019)	GDP (thousands $)	Area (sq miles)
Appalachia	241	11,547,242	499,766,765	97,495
Bayou	79	6,953,496	316,073,905	54,544
Bluegrass	293	15,256,804	702,684,032	144,978
Carolina	187	19,102,413	874,953,349	101,418
Central Plains	242	22,422,505	1,286,029,572	140,375
Chesapeake	123	41,692,558	3,071,833,670	39,250
Dixie	322	18,075,492	825,399,442	179,577
Erie	90	10,502,006	491,653,758	65,632
Everglade	31	16,400,151	752,134,747	31,158
Huron	182	24,597,271	1,219,102,584	88,099
New England	74	15,394,375	984,319,454	76,645
Northern Lakes	182	9,186,572	488,462,202	159,923
Totals	2,046	211,130,885	11,512,413,298	1,179,094

Sources: GDP, US Bureau of Economic Analysis; Population, Area, US Census Bureau

services that are actively supporting a change to an economy less impactful on planetary resources.

7.4 Participation in Governance

The original intent of this volume was to focus on the basic science underlying the design and implementation of an economy based primarily on biophysical principles. The exposition would be apolitical. In reality this intent proved to be naive. Biophysical analysis can help define the technical (e.g., energy, water, renewable resources, waste assimilation in the context of demand and supply) parameters of a sustainable economy. However, the body politic will largely determine if and how a biophysical economy is put in place.

The public, in a democratic political system, can facilitate (or thwart) achieving a biophysical economy through election of representatives at local, state, and federal levels who are in favor (or against) the required economic transition to a biophysical economy.

Note

1. A Greek word for a condition of happiness or well-being or flourishing – beyond mere sustainability. Interestingly, Wake Forest University (wfu.edu) advertises its Eudaimonia Institute.

8

Government

8.1 Introduction

The institutions and processes that broadly define *government* strongly influence the form and function of the current economy. There is every reason to believe that government will continue to be a determining factor in the development of a biophysical economy, or BPE. The role of government, however, might be fundamentally different than that played by the private sector. While the private sector will drive the transformation to a BPE, the public sector (i.e., government) will contribute "maps" and "rules of the road" to minimize any ruthless changes in economics wrought by departing neoclassical partisans. In the end, the creation and operation of a biophysical economy will of course be highly a political process. This process will be challenged by the degree of sophistication in understanding the complex nature of the required changes – environmental, social, economic – that legislators bring to the process.

The following discussion focuses on the United States federal government, but the overall concepts and suggestions concerning BPE might reasonably apply to state and local governments – perhaps even to other countries (e.g., Motesharrei et al. 2016). Extending a biophysical economy beyond the United States might, in fact, be a requisite for meaningful economic revision. Economic sustainability might have real meaning only when exercised at a global scale. If an economy aligned more closely with the principles of biophysics and recognition of limits to growth proves incompatible with the neoclassical model, countries that adopt biophysically based socioeconomic institutions and processes might suffer limitations imposed indirectly by remaining countries that continue to emphasize limitless growth.

The structure of government will influence its adaptability to support and participate in the implementation of a biophysical economy. High school civics class reminds us that the structure of the US federal government consists of three branches: executive, legislative, and judicial. Each will play a pivotable role in developing and operating a biophysical economy. The chapter is not intended as a review of a basic civics class, but simply as a parsimonious

DOI: 10.1201/9781003308416-10

introduction to how the US government might contribute to the development of a BPE.

8.2 Executive Branch

The executive branch enforces the laws. This branch of government comprises the offices of the president, vice president, and members of the cabinet. The executive branch includes key government agencies (Table 8.1). The listed constituent organizations attest to the breadth and potential reach of the executive branch in contributing to the development of a biophysical economy. It is difficult to identify dimensions of a complex socioeconomic and environmental system defined by biophysical economics that are not under the purview of one or more executive branch organizations. One advantage offered by these organizations in helping to set up a new economic model includes an ability to act with speed and efficiency, that is compared to the legislative branch. For example, executive orders enacted by the president are almost immediate. Another possible contribution is the incredible amount of data and information collected, managed, and distributed by organizations under the executive branch umbrella.

8.2.1 Independent Federal Agencies

The executive branch also includes independent federal agencies that operate with some degree of autonomy from the executive branch of government. Independent agencies comprise three main types: independent executive agencies, independent regulatory commissions, and government corporations. There appears to be no authoritative list of independent agencies within the US federal government. Table 8.2 includes independent agencies listed by the US Library of Congress (www.loc.gov). Again, a key challenge in using existing institutions in developing a biophysical economy will be to take advantage of existing structures and processes that appear consistent with the fundamentals of a new economic paradigm.

8.3 Legislative Branch

The US legislature consists of the Senate and House of Representatives (or Congress). The legislative branch is empowered to propose laws. It also confirms presidential nominations for heads of agencies and judges. The legislative branch importantly has the authority to declare war. While the preceding rather trite statements are clearly common knowledge within the US citizenry, they are offered again from the perspective of how current

TABLE 8.1

List of US Government Executive Agencies (www.loc.gov)

Departments	Constituent Organizations
Agriculture	Agriculture Research Service, Animal and Plant Health Inspection Service, Economic Research Service, Farm Service Agency, Forest Service, National Agricultural Library, National Resources Conservation Service, Rural Development
Commerce	Bureau of Economic Analysis, Census Bureau, International Trade Administration, NOAA Fisheries, National Institute of Standards and Technology, National Oceanic and Atmospheric Administration, National Ocean Service, National Technical Information Service, National Telecommunications and Information Administration, National Weather Service, Patent and Trademark Office
Defense	Air Force, Army, Defense Logistics Agency, Marines, National Security Agency, Navy
Education	Educational Resources Information Center, Institute of Education Sciences, National Library of Education
Energy	Lawrence Livermore National Laboratory, Los Alamos National Laboratory, National Nuclear Security Administration, Office of Science, Pantex Plant, Sandia National Laboratory, Savannah River Site, Southwestern Power Administration, Y-12 National Security Complex
Health and Human Services	National Institutes of Health, National Library of Medicine
Homeland Security	Citizenship and Immigration Services, Coast Guard, Federal Emergency Management Agency, Federal Law Enforcement Training Centers, Intelligence Careers, Secret Service
Housing and Urban Development	Government National Mortgage Association, Office of Lead Hazard Control and Healthy Homes, Public and Indian Housing
Interior	Bureau of Land Management, Bureau of Reclamation, Fish and Wildlife Service, Indian Affairs, National Park Service, Office of Surface Mining Reclamation and Enforcement, US Geological Survey
Justice	Bureau of Alcohol, Tobacco, Firearms, and Explosives; Drug Enforcement Administration; Federal Bureau of Investigation; Federal Bureau of Prisons; Office of Justice Programs; US Marshals Service
Labor	Bureau of Labor Statistics, Mine Safety and Health Administration, Occupational Safety and Health Administration
State	Bureau of International Security and Nonproliferation, Department of State Library
Transportation	Bureau of Transportation Statistics, Federal Aviation Administration
Treasury	Alcohol and Tobacco Tax and Trade Bureau, Bureau of Engraving and Printing, Bureau of Fiscal Service, Financial Crimes Enforcement Network, Internal Revenue Service, Office of the Comptroller of the Currency, United States Mint
Veterans Affairs	(None)

TABLE 8.2
List of US Independent Government Agencies

List of US Independent Government Agencies
Agency for Global Media
AmeriCorps
Central Intelligence Agency
Consumer Financial Protection Bureau
Environmental Protection Agency
Farm Credit Administration
Federal Deposit Insurance Corporation
Federal Housing Finance Agency
Federal Labor Relations Authority
General Services Administration
Institute of Museum and Library Services
National Aeronautics and Space Administration
National Archives
National Credit Union Administration
National Endowment for the Arts
National Endowment for the Humanities
National Railroad Passenger Corporation
National Science Foundation
Office of Government Ethics
Office of Personnel Management
Overseas Private Investment Corporation
Peace Corps
Pension Benefit Guaranty Corporation
Selective Service System
Small Business Administration
Social Security Administration
Tennessee Valley Authority
Thrift Saving Plan
United States Agency for International Development
United States Postal Service
United States Trade and Development Agency

governmental institutions might help assist in the design and execution of a biophysical economy.

The United States Senate is supported by 24 committees, many of which include a number of sub-committees (Table 8.3). A detailed description of the scope, jurisdiction, and function of each sub-committee lies beyond the intent of this chapter. The key point is that actions undertaken by the Senate in support of, or in challenge to, building and operating a biophysical economy will be largely impacted by the behind the curtain (to the interested public) activities of these sub-committees. The educational background, training, and experience evidenced by senators and, more importantly, by

TABLE 8.3

Senate Committees Supporting 117th United States Congress (www.congress.gov/committees)

Committee	Members/Sub-committees
Agriculture, Nutrition and Forestry	22/5
Appropriations	30/12
Armed Services	26/7
Banking, Housing, and Urban Affairs	24/5
Budget	22/0
Commerce, Science, and Transportation	28/7
Energy and Natural Resources	20/4
Environment and Public Works	20/4
Finance	28/6
Foreign Relations	22/7
Health, Education, Labor, and Pensions	22/3
Homeland Security and Governmental Affairs	14/3
Indian Affairs	12/0
Joint Committee on Printing	5/0
Joint Committee on Taxation	5/0
Joint Committee on the Library	5/0
Joint Economic Committee	10/0
Judiciary	22/8
Rules and Administration	18/0
Select Committee on Ethics	6/0
Select Committee on Intelligence	20/0
Small Business and Entrepreneurship	20/0
Special Committee on Aging	14/0
Veterans' Affairs	18/0

supporting staff members in relation to the many dimensions underlying biophysical economics will determine in no small part the efficacy of implementing such an economy.

The United States Congress (House) is supported by 24 committees and sub-committees (Table 8.4).

8.4 Judicial Branch

During the initial conceptualization of this book, the author naively took for granted that the execution of a major shift in economic paradigms would not suffer from judicial impediments. Of course, the judicial branch or court system in the United States includes the Supreme Court and federal judges that interpret the law, including its constitutionality, and apply the law in

TABLE 8.4

House Committees Supporting 117th United States Congress (www.congress.gov/committees)

Committee	Purpose	Members/Sub-committees
Agriculture		51/6
Appropriations		59/12
Armed Services		59/7
Budget		35/?
Education and Labor		51/?
Energy and Commerce		58/?
Ethics		10/?
Financial Services		55/8
Foreign Affairs		51/6
Homeland Security		35/6
House Administration		9/0
Judiciary		44/5
Natural Resources		47/6
Oversight and Reform		45/6
Rules		13/3
Science, Space, and Technology		41/5
Small Business		27/5
Transportation and Infrastructure		65/6
Veterans' Affairs		29/5
Ways and Means		42/6
Joint Committee on Printing		9/0
Joint Committee on Taxation		10/0
Joint Committee on the Library		9/0
Joint Economic Committee		10/0

individual cases. The judicial branch has the power to punish those found in violation of the law. Regarding violation of the law, it is imperative to determine whether the required changes in economic institutions and procedures to secure a biophysical economy would be legal. Certainly, this is no small undertaking.

The economic system that has evolved in this country over the past several hundred years has also benefited from (and co-generated) a corresponding monumental legal framework and history. Much of this derives from the Constitution. For example, the concept of private property is fundamental to capitalism and the Takings Clause of the Fifth Amendment reads as, "Nor shall private property be taken for public use, without just compensation." This clause intends to uphold the basic principle that government should not cause individuals to bear excessive burdens, even in support of ostensibly justifiable public works or public goods (Pilon 2017). This author claims no expertise or credentials in the interpretation of the constitutional law.

However, some even rudimentary background research (e.g., Treanor 1995) informs that America's founders equated private property as foundational to prosperity and freedom. The Constitution protects property rights – namely the rights of people to freely obtain, use, and dispose of private property (Pilon 2017). The origins of property rights in common law further emphasize the right of sole dominion, which includes the right to exclude others, a right against trespass, and a right of quiet enjoyment. The owner can freely exercise sole dominion to the extent that it does not infringe upon the property rights of others. If the implementation of a biophysical planned economy imposed limitations on an individual's property rights, for example, by potentially limiting the accumulation of personal wealth or restricting use of acquired property, would this violate the Fifth Amendment?

Often at issue in the context of the Takings Clause are the two associated public powers, namely the power of the government to secure property rights, or police power, and the power of eminent domain, where the government can take property for public use upon just compensation. Police power appears legitimately exercised when it is used to secure rights that provide certain public goods, for example, national defense or clean air (Pilon 2017). When the government acts through police power to stop the endangerment of others by an owner's activities, the owner is not entitled to compensation. For example, in relation to environmental quality, polluters need not be compensated for polluting. To what extent might the government employ police power to achieve the aims and goals of an economic revision to the nature and extent implied by biophysical economics?

Eminent domain is more of a regulatory power. This power provides government with an alternative process to secure property rights for public works – for example, building roads or creating habitat to protect wildlife. Eminent domain appears justified mainly in the context of enabling public projects without being held for ransom by individuals seeking more than just compensation. The power is "Pareto superior" in the vernacular of economists, that is at least one party (the public) is made better off, while no party is worse off, presuming just compensation is given (Pilon 2017). In relation to biophysical economics, when the government uses the power of eminent domain to take some rightful use by an owner to provide public goods, such as wildlife habitat, the owner must be compensated. The public must pay for the goods and services it wants (Pilon 2017).

This pedestrian description of the structure and function of the US government provides the basis for making several points concerning the role of government in the design and implementation of a biophysical economy.

8.5 Data Collection

Government agencies collect staggering amounts of diverse data and information that might be used to guide the design and implementation of a biophysical economy. Effectively all of the departments, agencies, and supporting committees listed in Tables 8.1 through 8.4 routinely collect data that are used to evaluate performance in relation to policy goals and objectives. These organizations collect data that quantify the status quo along the dimensions of their respective purview. The data can also be used to identify trends or make projections. Recognizing these information resources, a key activity to initiate the development of a biophysical economy will be a comprehensive survey across government to identify and access data streams that will be relevant. The data can remain decentralized in their locations for storage and management, dictated mainly by the collecting entity. However, a centralized data management system might benefit the effective utilization of these disparate data to help build and operate a biophysical economy.

An effectively managed biophysical economy will require data collected and made available in real time to evaluate the performance of the economy with respect to established management goals and objectives. One example of current capabilities in near real-time data applications is the US Forest Service early warning system, ForWarn II (Figure 8.1). This service provides updates every eight days on the status of forests and other land-use categories based on automated collection, analysis, and reporting technologies based on remotely sensed data (Figure 8.2). The application provides quantitative assessments of deviations in forest health (a greenness index) compared to established baseline conditions for different forest and land-cover types. The ForWarn II spans the United States, but because it uses satellite imagery, this technology could be expanded to a global scale.

The ForWarn II application is able to convey the implications of various environmental stressors on forest health. For example, the red colors in Figure 8.2 illustrate the locations and spatial extent of forest fires previous to the May 2024 reporting period. The effects of other stressors, including drought, disease, and insect pests, can be similarly assessed and visualized using the ForWArn II technology.

The ForWarn technology has been expanded to include the near real-time (8-day) assessment of agricultural productivity at a spatial resolution of 230 m^2 for more than 100 key crops grown throughout the United States (Konduri et al. 2020). The eight-day characterizations of crop productivity appear of substantial value in projecting harvests and pin-pointing specific crop and locations that appear highly productive, as well as identifying crops that are at risk from drought, disease, and other stressors. This technology would likely have additional importance in commodity markets and associated investments and trading.

The eight-day temporal resolution of these satellite-based technologies is rather the exception to most government monitoring programs. Other departments and agencies (e.g., USDA, NOAA, USEPA) aggressively collect data that could inform the routine operation of a biophysical economy, but the collections, analyses, and reporting typically occur on annual (or longer) timescales (e.g., US Census Bureau). As indicated throughout this volume,

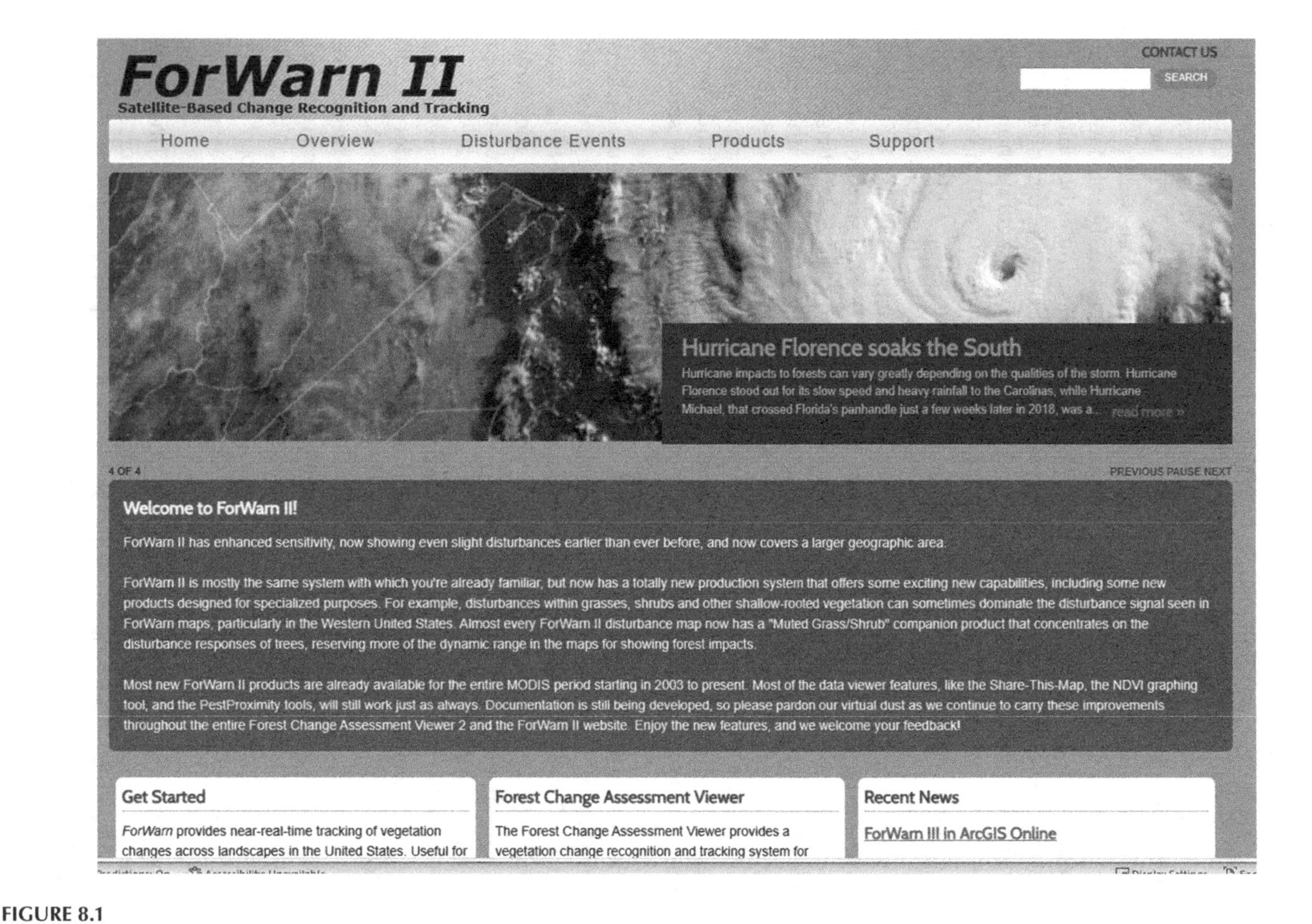

FIGURE 8.1
Home page for the US Forest Service early warning system. Credit: U.S. Department of Agriculture (https://forwarn.forestthreats.org).

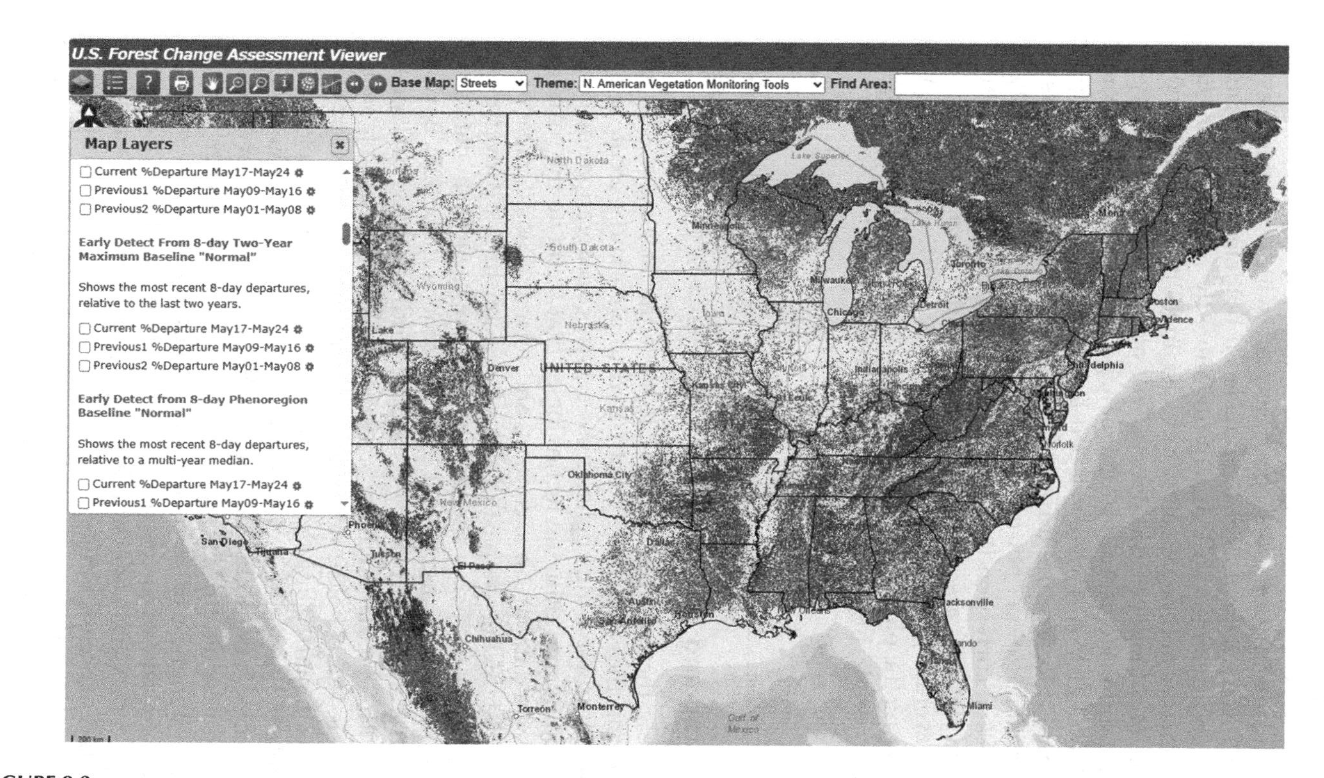

FIGURE 8.2
Remotely sensed departures of forest conditions compared to baseline for May 1–May 8, 2024. Credit: U.S. Department of Agriculture (https://forwarn.forestthreats.org).

characterizing and managing feedback among disparate socioeconomic-ecological structures and processes are key aspects of manifesting a sustainable economy. The corresponding challenge lies in making the government data and analysis available at timescales commensurate with the timescales of decision making. Management decisions made on outdated quantification of this complex and dynamic system will undoubtedly impact biophysical economic performance and negatively affect sustainability (Sterman 2000).

The operation of a biophysical economy will also require the ability to project future conditions to usefully guide management planning. For example, the USEPA and NOAA have developed technologies to forecast future changes in climate at local scales through the end of the 21st century. As an example, the Climate Explorer permits the user to specific a city or county in the United States and corresponding projections of a number of temperature and precipitation statistics are automatically provided. Figure 8.3 illustrates future values of average daily maximum temperature for Dane County, WI. The data underlying the plot can be easily downloaded for additional analysis and use, for example, in planning and decision making.

8.6 Data Sharing and Transparency

Key to establishing the required feedback loops is the sharing of data and information among those ultimately entrusted to plan and execute a biophysical economy. Data sharing implies cross-agency collaboration and cooperation in developing and utilizing technologies that facilitate the seamless transfer of disparate data streams, content, and format. Considerable progress has been made during the past decades in establishing E-Government data sharing protocols to ensure speed, accuracy, and security in data sharing (Bajaj and Ram 2007; Liu and Chetal 2005; Otajacques et al. 2007). Barriers remain largely in the form of privacy issues that address individual identities associated with data that might be accessed by NGOs, the private sector, and other interested stakeholders outside of government (Fedorowicz et al. 2010). The success in marshaling the largely unknown magnitude of government data to help in the design and execution of a biophysical economy will depend importantly on the ability to continue the evolution of e-government data sharing technologies and the successful navigation of barriers in technology and the willingness to share data in an efficient and secure manner.

Sharing large amounts of socioeconomic and environmental data across a diverse range of topics and content in essentially real time invites attempts from third parties to access, obtain, and further make the data available to consumers who might use them to their advantage or the disadvantage of the government. Cybersecurity will require continued attention and technological advances to reduce the risk of breaches in data security as complex

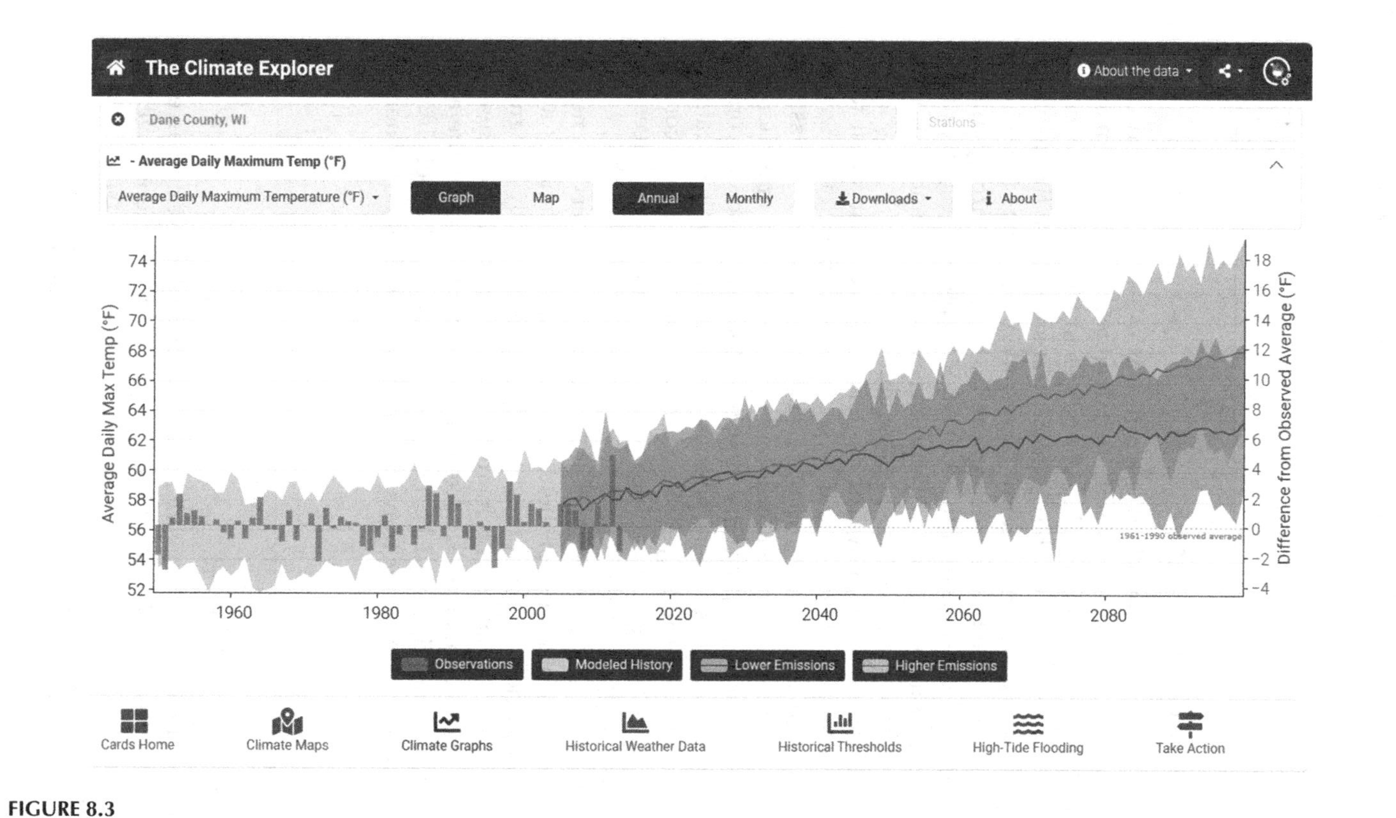

FIGURE 8.3
Example projection of average daily maximum temperature for Dane County, WI, for alternative low and high emissions climate change scenarios. Credit: U.S. Climate Resilience Toolkit Climate Explorer (https://crt-climate-explorer.nemac.org).

e-communications expand in the building and operation of a biophysical economy.

8.7 Government and Finance

As previously suggested, the role of government in the design and establishment of a biophysical economy will include government investments in the form of direct spending as well as incentives in the form of tax credits and deductions associated with the economic transition from the neoclassical paradigm to a sustainable economy. Government workings in the realm of financing an economic transformation can be facilitated through macroeconomic analyses framed within the context of biophysics (e.g., Jacques et al. 2023). The following Chapter 9 will further explore the contributions of government in financing a biophysical economy.

The transformation to a biophysical economy will come at a cost to the government. Funding will be required to set up government organizations and processes to collate, manage, analyze, and otherwise use the vast data collection activities that can usefully inform the design and operation of a biophysical economy. New programs that monitor the pulse of economic activity in response to government policies aimed at economic sustainability will need sustained funding. Potential reductions in annual economic growth might reduce government income in the form of taxes, fees, and other sources of revenue.

At the same time, the advances of new technologies and services needed to perform the transformation of the current neoclassical economy to sustainability will add new sources of revenue. Society is currently recognizing the financial gains associated with increased employment in the renewable energy sector and technologies directed at removing CO_2 from the atmosphere. Careful accounting will be necessary to assess the net financial impacts of the needed economic transformation to a biophysical economy.

8.8 Government and Policy

The government touches nearly every imaginable aspect of the human enterprise as underscored by the breadth and depth of departments, agencies, and supporting committees listed in Tables 8.1–8.4. This diversity and complexity in governmental organization also provides a platform for government to reinvent its role in stimulating an economic transformation towards sustainability (Costanza 2010). Government policies that focus on creating an economy aimed at increasing human well-being and quality of life, as opposed (or in addition) to continued increases in material consumption, can contribute to the redesign of society that increasingly values a sustainable economic future. Government provides the organization at scale that can conduct the cross-sector, multi-dimensional integration of disparate socio-economic and environmental policy dimensions with a focus on increased well-being for humans and their biophysical life support systems (Wang et al. 2022).

9

Finance

9.1 Introduction

Obviously, finance is a critical component in the design and implementation of an economy based on biophysics. The contributions of the financial sector appear of no less importance than the policy and decision-making roles of the private sector, the general public, and government in forming and enacting a biophysical economy. The roles of finance in relation to a biophysical economy are multifaceted. For example, failure to transition to a sustainable economic model will incur costs in the form of continued degradation of the environment with associated costs and financial damages to infrastructure resulting from increased frequency and intensity of natural disasters, including fires, floods, drought, and intense storms. Patterns in lending and investment will undoubtedly change in relation to increased emphasis on truly sustainable development. Transition to an economy in concert with potential limitations imposed by finite planetary life-support systems might well impact the nature and performance of the stock and commodity markets.

The US financial sector is dominated by the US Federal Reserve system, including 12 federal reserve banks. Thousands of private banks and other financial institutions make up the bulk of the private financial sector. The various key equity exchanges also play pivotal roles in the trading of stocks, bonds, and other financial instruments in a complex and dynamic economy. Commodity markets are also essential to a functioning economy. All of these financial organizations and metrics will play important roles in the translation of the current economy to one that emphasizes sustainability.

The following chapter highlights some of the financial challenges that will have to be addressed during the transition from the current neoclassical system to an economy based on biophysics and directed towards sustainability.

DOI: 10.1201/9781003308416-11

9.2 Financial Impacts of Environmental Degradation

Patching and propping up the current growth-focused economic model will continue to incur the costs associated with continued environmental degradation and related impacts on human health, the environment, and built infrastructure – not to mention social and natural capital (Costanza 2010).

9.2.1 Air Pollution

The US Environmental Protection Agency routinely uses its Environmental Benefits Mapping and Analysis Program-Community Edition (BenMAP-CE) to estimate the health-care costs of air pollution (Birnbaum et al. 2020). The default BenMAP-CE considers only the costs associated with hospital and emergency department admissions. Birnbaum et al. (2020) expanded the analysis to include a more complete accounting of ambulatory and costs through analysis of health insurance claims in addition to work loss costs. The net result was that an additional 40% of health-care costs could be justifiably linked to the health impacts of air pollution. Related studies concluded that the Clean Air Act, for example, reduced the costs of respiratory and cardiovascular hospitalizations by $488 million (in 2006 $USD) (Birnbaum et al. 2020).

The economic costs of air pollution extend beyond human health (Franchini et al. 2015). Important non-health-related costs of air pollution include visual intrusion and loss of tourism, lower land property values, and damage to buildings and agricultural crops. Deluccchi (2004), for example, performed a comprehensive assessment and estimated that visibility-related damages ranged from $37–45 billion/year, building damages were $0.4–8 billion, and damages to crops were valued at $2–6 billion for the 1990–1991 period.

In a novel study, Eckelman et al. (2020) examined the greenhouse gas emissions produced by the health-care industry in response to increased access to the system necessitated by increased air pollution. Using a state-by-state analysis based on environmentally extended input-output model the authors estimated health-care-associated GHG emissions rose 6% from 2010 to 2018 to 1,692 kg per capita in 2018. This corresponded to a loss of 388,000 disability-adjusted life years. Eckelman et al. (2020) suggested that the health-care industry undertake efforts to reduce consumption of resources, decarbonize power generation, and invest in preventative care to reduce the overall air pollution footprint of the health-care industry.

It is possible to adopt a broader perspective on the impacts of the economy and estimate the gross external damage (GED) associated with emissions from specific industries and activities as part of a national accounting of nation-scale economic activity (Muller et al. 2011). The GED can also be compared with value added (VA) to the economy. In relation to developing a sustainable economy, the objective would be to minimize the overall GED for an emitter and similarly minimize the GED/VA ratio, where a ratio >1 identifies an industry or business activity whose value added to the economy fails to outweigh the damages it produces. Based on their detailed analysis, Muller et al. (2011) estimated GED in billions of $USD per year (2002 prices) for a substantial number of business sectors, with several of the highest noted below:

- 62.6 for utilities (0.34 GED/VA)
- 32.0 for agriculture and forestry (0.38)
- 26.4 for manufacturing (0.01)
- 23.2 for transportation (0.10)
- 14.7 for construction (0.03)
- 10.7 for waste management services (0.08)

The study concluded that $184 billion in overall GED across the sectors that were analyzed. Drilling down into specific industries, Muller et al. (2011) estimated the greatest GED of $53.4 billion for coal-fired power generation, which demonstrated a corresponding GED/VA of 2.20.

The GED and VA metrics could contribute to the continued evaluation of economic activities during the transformation to a biophysical economy.

9.2.2 Water Pollution

The impacts of water quality violations as a result of increased pollution have measurable costs on human health (Alzahrani et al. 2019). The authors performed an extensive state-by-state empirical analysis of the implications of water quality on related health-care expenditures. Based on the results of multi-factor regressions, it was estimated that the per capita savings in health-care expenses per a 1% reduction in violations of water quality standards was $0.32 for the state and $0.35 per capita for the surrounding three states throughout the US study domain. These small reductions translate, for example, to a $1.2 million projected decrease in health-care expenditures for the state of Oklahoma. Its surrounding three states (Texas, Arkansas, and Kansa) would save approximately $71.1 million in annual costs. Propagating the results nationwide indicate a $98.9 million in-state and $698.2 million regional annual savings for each 1% reduction in violations of water quality standards (e.g., under the Clean Water Act) (Alzahrani et al. 2019).

An ever-expanding economy imposes additional water-related costs. For example, the impacts of agricultural production on water resources were estimated for treatment of surface waters for microbial pathogens and infrastructure costs to treat nitrate and pesticides for the year 2002 (Tegtmeier and Duffy 2004). The annual costs for treating pathogens was $118.6 million. Costs for nitrate infrastructure were $188.9 million and $111.9 million was spent for pesticide treatment – for a 2002 total of $419.4 million.

9.2.3 Ecosystem Services

The financial impacts of continued environmental degradation can also be measured as a loss in ecosystem goods and services as defined by the Millenium Assessment (MEA 2005) and others. Comprehensive assessments of the monetary value of ecosystem goods and services suggest that they far exceed more conventional measures of global GDP (e.g., Costanza et al. 1997).

Payments for ecosystem services (PES), or incentives for environmental protection, have been in place for several decades, but they have gained increasing recognition in mainstream economics in relation to increased awareness and valuation of ecosystem services (Friess et al 2015; TEEB 2010). PES projects have been developed or are in various stages of implementation worldwide (Table 9.1).

Recent advances in valuing ecosystem goods and services have focused on their time-value (Drupp et al. 2024). These authors determined that as ecosystem services decline relative to continued increases in per capita GNP, that price adjustment factors be factored into the valuation (e.g., willingness to pay models) of ecosystem services. Declines of ecosystem services in absolute terms (e.g., coral bleaching, deforestation) that generate scarcity in ecosystem services should be treated similarly in their valuation as standard practice in cost-benefit analysis used to support economic decision making.

9.3 Financial Impacts of Natural Disasters

Increasing frequency and severity of natural disasters associated with global climate change portend increasing financial risks (Chang and Zhang 2020; Ratcliffe et al. 2020). Putting in place an economic model based on biophysics might begin to alleviate the financial costs of recurring natural disasters. Data from NOAA demonstrated a general increase in the number of billion-dollar natural disasters since 1980 (Table 9.2). NOAA tallied 376 such disasters from 1980 through 2023 with a total cost of $2.67 trillion dollars, or approximately $61 billion per year over that period. Annual costs of natural disasters have been steadily increasing over the past 43 years to more than $120 billion. It is recognized that these increases reflect, in part, the overall increase in value of infrastructure, produced by a growing economy, that are potentially impacted by natural disasters, as well as an increase in the number of yearly disasters. Even more important, these monetary losses to public and private infrastructure do not include the estimated 16,350 lives lost to natural disasters since 1980.

The implicit hypothesis is that, in the longer term, the frequency and intensity of natural disasters would diminish in response to a transition to an

TABLE 9.1

Payments for Ecosystem Services Summarized from Friess et al. (2015)

Ecosystem	Service	Description	Reference
Agricultural	water quality	New York (USA) agreement to modify upriver farming practices as alternative to new downriver water treatment facilities	Rosa et al. 2004
Rainforest	carbon, biodiversity	Protection of carbon stocks and biodiversity in Lacandon (Mexico) rainforest fragments	Kosoy et al. 2008
Rainforest	carbon	Juma (Brazil) REDD certified project to avoid deforestation and carbon emissions	Borner et al. 2013
Coastal ocean	fisheries	European Union invests to protect Mauritania fisheries from over-exploitation	Binnet et al. 2013
Rainforest	biodiversity	Payments to avoid deforestation in Rwanda lower montane forest, reduce hunting, and reduce mining	Gross-Camp et al. 2012
Mangrove	carbon	Conserving and restoring mangroves in Gazi Bay (Kenya) to capture 300 tonnes of C per year	EAFPES 2013
Agriculture, reef	multiple	Co-payments to reduce sediment and nutrient loading to Great Barrier Reef (Australia)	Brodie 2014
Rainforest	water quality	Steel company paying communities to protect forests in western Java (Indonesia)	Mbak 2010
Coral reef	multiple	PES under development to protect marine biodiversity in the Philippines	CTI 2012
Dry forest	carbon	REDD+ project to sequester 7.1 million tons over 30 years in Cambodian dry forest	Forest Trends
Agricultural	water quality	French farmers compensated to reduce nitrate contamination in Vittel catchment	Perrot-Maitre 2006
Agricultural	biodiversity, water quality	UK farmers paid to reduce livestock densities and fertilizer use	Dobbs and Pretty 2008

economic system more compatible with biophysics and Earth's life support systems. In the energy transition implied by a sustainable economy, GHG emissions would be reduced in absolute terms to produce measurable reductions in global atmospheric CO_2 concentrations from the current (at the time of submitting this manuscript) ~420 ppm. Each decrease in 1 ppm of atmospheric CO_2 requires the removal of about 8 billion metric tons of CO_2 from the global active carbon pools (source: Carbon Dioxide Information and

TABLE 9.2

Time Comparisons of Billion-dollar Natural Disasters in the United States (from NOAA)

Period	Number Events	Events/Year	Cost ($ billion)	Cost/year ($ billion)	Deaths
1980–1989	33	3.3	213.6	21.4	2,994
1990–1999	57	5.7	326.8	32.7	3,075
2000–2009	67	6.7	604.2	60.4	3,102
2010–2019	131	13.1	967.5	96.8	5,227
2019–2023	102	20.4	603.1	120.6	1,996
All years	376	8.5	2,661.1	60.5	16,350

TABLE 9.3

Billion-dollar Natural Disasters in the United States (from NOAA)

Disaster Type	Number Events	Frequency (%)	Total cost ($ billion)	Cost/event ($ billion)	Deaths
Drought	31	8.2	352.9	11.4	4,522
Floods	44	11.7	196.6	4.5	738
Freeze	9	2.4	36.4	4.0	162
Severe storm	186	49.5	455.2	2.4	2,094
Cyclone	62	16.5	1,379.3	22.2	6,897
Wildfire	22	5.9	142.4	6.5	535
Winter storm	22	5.9	98.3	4.5	1,402
All disasters	376	8.5	2,661.1	7.1	16,350

Analysis Center, Oak Ridge National Laboratory). Longer-term reductions in atmospheric CO_2 emissions would hopefully result in fewer, less intense, and less costly natural disasters. It is recognized that this desired result, if it occurs, would likely manifest over decades to centuries, given the natural residence time of carbon dioxide and other GHGs in the atmosphere.

NOAA further dissected the billion-dollar natural disasters from 1980 to 2023 to specific types (Table 9.3). Severe storms were the most frequent among the seven categories defined by the NOAA data accounted for about half of the disasters and about $455 billion in damages. Tropical cyclones (hurricanes) represented only about 17% of the billion-dollar events, but they proved the most costly at $1.38 trillion. Floods and drought accounted for nearly 20% of the billion-dollar events and cost approximately $550 billion in damages. Winter storms and freezes tallied more than 8% of the events with combined damages of $135 billion. Wildfires represented 6% of the 376 recorded events, but cost $142 billion.

Of the 16,350 reported deaths associated with the 376 billion-dollar events, nearly seven thousand people were killed by tropical cyclones (Table 9.3). Somewhat surprising was that droughts accounted for more casualties, 4,522 than severe storms, 2094 – more than a twofold difference. Winter storms and freezes accounted for 1,564 deaths.

Establishing a biophysical economy does not portend that the frequency and intensity of billion-dollar natural disasters will immediately decline. The presumptions are twofold: One, the transition to an energy technology portfolio that emphasizes renewable forms of energy (e.g., wind, solar, hydropower) will reduce GHG emissions and reverse the trend of an increasing number of annual billion-dollar events. Shifting to an economy that is not locked into continuous growth in absolute terms can also help reduce the growth in GHG emissions and indirectly contribute to reversing the trend, at least since 1980, of these costly disasters.

Two, the frequency and intensity of natural disasters influenced by climate change are symptoms of a larger-scale challenges to planetary life support systems posed by population size and associated demands for resources, particularly those demands imposed by highly industrialized and technologically developed societies (Wackernagle and Reese 1996; Reese 1992).

9.4 Lending

The successful establishment of a sustainable economy will depend on changes in patterns of lending by banks and other financial institutions to move away from investment portfolios encumbered by investments in unsustainable activities and potentially unacceptable risks (e.g., the mortgage finance crisis in 2008).

Increasingly, banks are exercising due diligence in evaluating projects in relation to sustainability. For example, Deutsche Bank (2023) has established a framework to evaluate its lending activities in relation to a transition to low carbon and sustainable business models. The framework characterizes sustainable financing according to three parameters: use of proceeds, company profile, and sustainability-linked solutions. The framework is used to evaluate client proposals in relation to the three parameters. If any one of the three parameters is met, the proposal is eligible for classification as sustainable (Deutsche Bank 2023).

Parameter 1 – Use of Proceeds

Two basic considerations are used to evaluate the use of proceeds in relation to sustainability. One, if the proposed activities to be financed can be demonstrated to sustain, improve, and protect the environment, the proposal will be classified as consistent with Parameter 1. Two, if the proposed activities clearly enable social development, especially for marginalized socioeconomic groups, but also for the general public, Parameter 1 will be satisfied.

If the proposed activity to be financed fails to satisfy Parameter 1, the activity will be reviewed in relation to Parameter 2.

Parameter 2 – Company Profile

The financing is eligible of the company derives at least 90% of its revenues from environmentally and/or socially sustainable activities and such activities can be identified for one or more of the company's activities. Deutsche Bank (2024) has accordingly identified environmentally six general areas of sustainable activities that will be considered in considering client eligibility and evaluating financing according to Parameter 2 (Table 9.4).

Similarly, the bank has developed a classification of activities that it considers socially sustainable in relation to meeting Parameter 2 requirements (Table 9.5).

Parameter 2 can also be satisfied if the proposing entity is not involved in activities excluded according to Deutsche Bank criteria.

TABLE 9.4

Environmentally Sustainable Activities Defined by Deutsche Bank (2023)

Climate change adaptation: activities that substantially reduce GHG emissions or increase energy efficiency
Climate change adaptation: measures that adapt to physical risks caused by or intensified by climate change
Sustainable use of water and marine resources: protection of marine resources and terrestrial resources, including water, high-carbon stock ecosystems, and other primary resources
Transition to a circular economy: prevention of waste and promotion of recycling and reusing materials
Pollution prevention: reducing pollution and reduction in general resource utilization
Protection and restoration of ecosystems and biodiversity

TABLE 9.5

Socially sustainable guiding objectives defined by Deutsche Bank (2023)

Basic infrastructure: enablement of basic human rights, e.g., food, clean water, sanitation, labor protection, transportation, and energy
Essential services: access to health care, education, and financial services
Housing: access to or building affordable housing
Financing: equal access to banking and financial services; advisory services to micro, small, and medium-sized enterprises
Employment: prevention of unemployment caused by socioeconomic crises
Secure food systems: access to safe, nutritious, and sufficient food; resilient agricultural practices; reduced food loss and waste; improved productivity of small-scale producers
Socioeconomic: access and control of assets, services, resources, and opportunities; equitable participation in markets and society; reduced income inequality

If the proposed activity to be financed fails to satisfy Parameter 2, the activity will be reviewed in relation to Parameter 3.

Parameter 3 – Sustainability-Linked Solutions

Deutsche Bank will support sustainability-linked solutions that provide incentives for clients that achieve predetermined sustainability performance targets (SPTs). The SPTs requirements established by the bank are based on recognized industry standards (Table 9.6). The SPTs are intended to be ambitious, consistent, and material to the client's core business and core economic activities. SPTs must be linked to key performance indicators that are verifiable and reportable.

The bank has developed a formalized and robust set of governance processes to ensure that transactions and financial products classified as sustainable comply with the Framework.

9.5 Investment

Investments involve making decisions concerning multiple attributes in an attempt to balance risks and rewards. Historically, the risks and rewards have been defined and measured in financial terms (e.g., avoided costs, shared profit). More recently, risks and rewards have expanded to include measures of environment, social, and governance (ESG). Conventionally, the balancing in investment takes the form of allocating financial resources to a portfolio that includes a range of economic sectors and classes of assets (deLlano-Paz et al. 2017; Bar-Lev and Katz 1976).

One somewhat counterintuitive impact of the financial systems on the economy occurs when financial instruments chase one another – money

TABLE 9.6

Industry Guides for Sustainable Performance Targets (SPTs) Identified by Deutsche Bank (2023)

LMA/LSTA/APLMA Sustainability-Linked Loan Principles
ICMA Sustainability-Linked Bond Principles
ISDA Guidance for Sustainability-Linked Derivatives
LMA – Loan Market Association
LSTA – Loan Syndications and Trading Association
APLMA – Asia Pacific Loan Market Association
ICMA – International Capital Markets Association
ISDA – International Swaps and Derivatives Association

begets money; it becomes more profitable for a company to simply invest rather than produce its usual goods or services (Moore 2015). This behavior can lead to under-investment in capitalization and shortfalls in the production of necessary commodities – leading to scarcity and/or high prices. For example, over-investment in financial instruments and corresponding under-investment in exploration and extraction puts additional pressure on replenishing depleted stocks of fossil fuels, particularly crude oil.

9.6 Biophysical Economy and Equities

An important financial consideration affecting the design and implementation of a BPE concerns equity markets. Investments in stocks, bonds, and other similar instruments are largely based on anticipated economic growth and shared profits. If economic growth would be regulated by biophysical constraints and consequent regulations, a fair question is: what impact might a BPE have on the stock market and similar investments? Put bluntly, would an operating biophysical economy tank the stock market?

Clearly the potential impacts of a biophysical economy on liquid investments are of paramount importance given the significance of these kinds of financial instruments in retirement portfolios for millions of people. The transformation to a sustainable economy cannot pose unacceptable risks to individuals who have spent a lifetime saving and investing for retirement or other uses. The importance of IRAs and 401-K plans in corporate benefits packages in attracting and retaining highly skilled and experienced employees cannot be ignored in the planning and execution of the envisioned economic transformation to a sustainable economy. It could be argued that the lack of sustainability in the current economic system will eventually take its toll on investments in the longer term – maybe around 2040 if the Meadows et al. (1972) projections come to pass. However, the socioeconomic and political fallout from imminent retirees encumbering the major portion of risk in the short term might prove sufficiently destabilizing to current government and private sector institutions to require avoiding such circumstances even at the delay of implementing a sustainable economy.

There might be reason to believe that a biophysical economy might produce positive impacts on equity markets. In addition to the importance of anticipated economic growth in informing investments in equities, it is common knowledge among investors that the market generally responds negatively to uncertainty. Reduced opportunities for continued high (unsustainable) growth and the associated disaffection for liquid investments might be offset, at least to some extent, by increasingly reliable returns from a sustainable economy, even if rates of return are lower in absolute terms.

9.7 Costs of a Biophysical Economy

The benefits of a biophysical economy on national sustainability appear rather self-evident and are discussed elsewhere in this volume. However, the design and operation of such an economy do not come free of charge. Reasonable and important questions would inquire about the costs of initial implementation and day-to-day function of such an economy. Identifying the sources of funding a biophysical economy seems necessary as well.

9.7.1 Design

A biophysical economy is, by definition, a highly organized and planned economy tied mechanistically to energy, renewable and nonrenewable resources, and waste assimilation. The design of such an economy – or developing the plan – as a transition from the current neoclassical dogma will require funding. At this point, it remains unclear how such a design could materialize or who might demonstrate the authority to create an initial design. The conversation concerning the efficacy of a biophysical economy has yet to progress to a level of importance among most mainstream financial institutions.

9.7.2 Establishment

Setting up initially the new institutions and processes required to implement the plan and complete the transition to a biophysical economy will require funds. At the very least, as suggested earlier in the discussion, an important component will include the ability to collate and evaluate the diverse data streams provided by the government (and presumably other sources).

9.7.3 Operating Costs

The day-to-day operations of the newly established biophysical economy will also incur costs. It appears reasonable to anticipate that the funding for the transition to a biophysical economy and its continued (sustainable) operations will come from a variety of sources. Government entities can propose biophysical economy operating funds as part of annual budgets in the government planning cycle. Private financial institutions can build the operating costs of operating a biophysical economy into the prices of goods and services that ultimately pass to the consumer – consumers benefit from the comparative stability and predictability of a sustainable economy.

Fundamental to the operation of a biophysical economy are the costs of monitoring energy and resource use and analyzing these data in relation to the availability of corresponding resources. Many of these kinds of likely useful data, particularly for natural resources, agriculture, energy, and economic metrics, are already being collected by various government agencies (as outlined in Chapter 8). Current and projected budgets for the participating agencies data collection services might provide some indication of the costs of utilizing this information and performing the feedback analysis necessary to evaluate the performance (metrics) of a biophysical economy in relation to the operative economic planning goals and objectives.

Sustainable operations of private corporations in support of a biophysical economy will require these private sector organizations to allocate funds internally to define and achieve sustainable development goals. Financial resources will also be required to put in place and monitor for performance the business structures and processes necessary to support corporate sustainability programs (i.e., ESG).

10

Policy, Society, and Culture

10.1 Introduction

Policy-driven boundaries are likely as important as biophysical planetary boundaries in designing and establishing a biophysical economy. In this discussion, policy refers, in the meaning of Mirriam-Webster, to a defined course of action selected among alternatives in the context of historical or current conditions that guides decision making now and in the future. Policy often refers to a strategic plan or a system of guidelines directed towards shared goals or objectives and conducted using accepted procedures, particularly by a body of government. The main difference being that policy is invented by humans and can be changed by humans; planetary boundaries are determined by laws of physics, chemistry, biology, and geology and are not negotiable – regardless of the underlying operative economic paradigm, neoclassical or biophysical economy.

Scientifically informed policy can help define the boundaries (e.g., legislation) within which other agencies (e.g., private business, financial institutions, NGOs) can work collectively to sustain economic activity through the implementation of a biophysical economy. Policies are developed and executed at multiple scales and levels of organization. Developing and coordinating policies that facilitate the establishment of a biophysical economy in a consistent manner across local, state, and federal jurisdictions will be necessary.

10.2 Policy and Economic Worldview

"Economic ideas matter," proclaimed Beinhocker (2012), who underscored the continued influence of the writings of Adam Smith, now two centuries ago, on how current powers in government, business, and the media think about markets, regulations, and the role of the state. The 19th-century words of Karl Marx appear to have instigated revolutions in various parts of the globe and helped lay the ideological foundations of the Cold War. Milton

DOI: 10.1201/9781003308416-12

Friedman set the stage for Reaganomics (Beinhocker 2012). Historical experience more than suggests that changes in economic thinking will change politics and policy as well (Table 10.1). The design and operation of a biophysical economy will not only require radical policy directives to come to fruition, but also will, in turn, change politics and policy going forward (Ji and Luo 2020).

10.3 Policy and Politics

Economic policy is not made in a political vacuum and economic advice that ignores politics does so ill-advisedly (Acemoglu and Robinson 2013). Accordingly, economic analysis needs to characterize conditions where economics and politics will likely run into conflict and develop policy that attempts to account for the conflict and anticipates potential political backlash. Labor unions provide a good example. Economists would strive to limit or remove the ability of a union to exercise monopoly power and raise wages for its members. Yet labor unions do not only influence the function of the labor market. Historically, unions have contributed to the formation or preservation of democracies (Acemoglu and Robinson 2013). Reducing the power of unions to affect wages can lead directly to de-unionization and reduced market power. These circumstances can indirectly tilt power to otherwise dominant groups and interests in society that are not necessarily advocates of democratic policies and institutions.

TABLE 10.1

Worldview, Scope, and Methods of Different Forms of Economics (Ji and Luo 2020)

Economics	Position of the Economy	Scope	Methods
Classical	Portion of human society that provides welfare	• How the economy provides daily bread • Interaction with the ecosystem (Physiocrats) and society (Political Economy)	Historical and philosophical analysis
Neoclassical	Whole of the real world	• Nominally the whole world • Only the monetized portion	Quantitative and mathematical analysis of monetized values
Ecological	Subsystem of the ecosystem	• Economic process as biophysical process • Extracting material and energy to fuel human society • Effects of process on sustainability of ecosystem and society	Biophysical laws, multidisciplinary studies

10.4 Society and Culture

Humans are fundamentally social beings. Society refers to a broad group of people who live in a definable territory and share common social organizations or institutions. Territory can range from neighborhood to nation. Corresponding culture includes the beliefs, values, norms, practices, and artifacts of the society. Society and culture largely define each other (Little 2023). At least three aspects of culture are relevant in originating a sustainable economy:

- Almost all human behavior is learned. This assertion provides hope for the ability of people to make changes that will be required to achieve sustainability (e.g., reduced consumption, focus on efficiencies, reduced waste).
- Culture is innovative. Different societies are challenged by similar problems, but the solutions can be quite different. The overall road to a sustainable economy will comprise many separate pathways that provide innovative and culture-specific solutions to common sustainability challenges (e.g., amount, availability, and quality of energy, water, and food).
- Culture is restraining. Interpersonal feedback that imposes some degree of decorum among people as they interact also provides for some optimism that an economic transition can occur without large-scale socioeconomic chaos.

Relevant to the necessary economic transition, humans are slow to change values systems and associated activities (Page 2005). This characteristic social inertia suggests limitations to the timeliness and overall success of "bottom-up" approaches in effecting sweeping, near-term changes in society, its institutions, and its artifacts directed at economic sustainability. Similarly, the US government – as the collected effort of humans in a democracy – demonstrates a corresponding lethargy in making large-scale socioeconomic changes over the short term, although it possesses several mechanisms (e.g., Executive Orders). Thus, "top-down" approaches to implement a major economic transition appear similarly constrained.

10.5 Globalism

This book has focused mainly on the possibility and mechanism for building and operating a sustainable economy at the scale of the United States.

Ultimately, this economic transition must transcend all political borders and geographies to achieve meaningful sustainability at a planetary scale. Think globally, act globally.

Global problems require global solutions (Waring et al. 2023). For example, lines that define political boundaries at various scales, ranging from local townships to nations, drawn on a map are irrelevant to patterns of physical global atmospheric circulation that translate local emissions of greenhouses gases (GHGs) to increased global atmospheric concentrations and associated impacts of climate change on humans and their infrastructure – which manifest somewhat ironically at local scales. Of course, tangible and measurable actions undertaken by humans to address global issues necessarily occur locally – clearly, any global reach results from the collective local efforts undertaken at more localized scales. In keeping with the GHG example, there is no physical reason to anticipate that local efforts to reduce emissions will provide corresponding local relief from climate change impacts. Any absolute reductions in local emissions are diluted through regional and global patterns of atmospheric circulation. Correspondingly, one challenge in putting a biophysical system into place lies in convincing local participation with geographical sufficiency to obtain economic transition at the global scale.

The ethnosphere represents the sum total of all cultures on Earth (Little 2023; Rathje 2009). The ethnosphere is the collective cultural heritage of the human species – the ways of thinking, being, and orienting on the planet. Importantly, humans have adapted to historical changes in the conditions of existence through diverse cultural inventions. The invention of a new economic paradigm that is compatible with Earth's biophysical life-support systems presents perhaps one of the greatest cultural challenges ever faced by humans. Participation, agreement, acceptance, and implementation across multiple diverse cultures in changing the fundamentals of systemic economic interactions are also opportunities to alter the course of human inhabitation of the planet towards a sustainable future. The ethnosphere, in the end, emerges as the fundamental cultural unit that will determine the success in building and operating an economy that is globally sustainable.

Claims that a biophysical-based economy must ultimately be global in scale to achieve sustainability are not intended to imply or recommend global governance (e.g., Bierman et al. 2012; Ivanova 2011; Goldin and Vogel 2010; Rosenau 1995). Global communities have become increasingly compressed in space and time, as evidenced, for example, by the rapid spread of recent pandemics (e.g., COVID-19) and the evolution of real-time global financial transactions. A global government might prove to be the most efficient and effective way to implement the necessary transition towards a sustainable economy (and society). However, thousands of years of observations suggest that *H. sapiens* is not sufficiently advanced in its evolution to working on its own behalf – even to the point of perhaps reducing risk of extinction (Waring et al. 2023). Organisms, including humans, routinely take the lives of other organisms to secure scarce resources (e.g., food, territory, money). Humans, however, appear to be the only organisms on the planet that kill one another over disagreement on ideas – mental fabrications. Artifacts of cultures. Global governance is not likely in the future of this species.

10.6 Policy and Models

The simultaneous consideration of socioeconomic, political, and environmental components critical to the design and implementation of a biophysical-based economy emerges as a key challenge (Costanza et al. 2007a; Diamond 2005). Large-scale, integrated global modeling systems, in the form of earth system models (ESMs) or integrated assessment models (IAMs), can play an important role in helping to understand the complex interrelations among diverse human institutions and biogeochemical systems as they potentially impact sustainability and supporting policymaking (Edwards 1996). Earth system models emphasize the natural science system components of global climate change and strive for understanding climate change and predicting future climates with increasing accuracy as the models develop. The integrated assessment models, in contrast, use aggregated outputs of ESMs and add the anthropogenic socioeconomic and political structures and processes to help understand the impacts of climate change on the human enterprise and evaluate the benefits and costs of mitigating change.

IAMs are of principal interest in developing this chapter. One such IAM is the International Futures (IFs) model (e.g., Hughes, 2019, 2009). There are, of course, others (e.g., Costanza et al. 2007b). Figure 10.1 illustrates the key structural elements and connecting processes addressed by the IFs model.

The IFs model was developed to help understand the interrelationships among the components and processes of a complex simulated socioeconomic–environmental system as they influence the outcomes of base case and user-defined planning scenarios on human development, social change, and environmental sustainability (Hughes 2009). The IFs model permits the evaluation of the implications of user-defined policy and planning scenarios as they might impact modeled structural components of education, health, population demographics, infrastructure, agriculture, technology, energy, economic development, and environmental resource quality and availability (Figure 10.1).

The intent here is not to provide an in-depth description of all the details represented in the IFs. The interested reader is referred to Hughes (2019, 2009). However, examples of process-level interactions among the modeled structural elements include labor, income, demand, supply, prices, investment, efficiencies, water, and resource use as they influence the economy. Mortality, fertility, income, government expenditures, and other factors determine, in part, responses in education, health, and overall population.

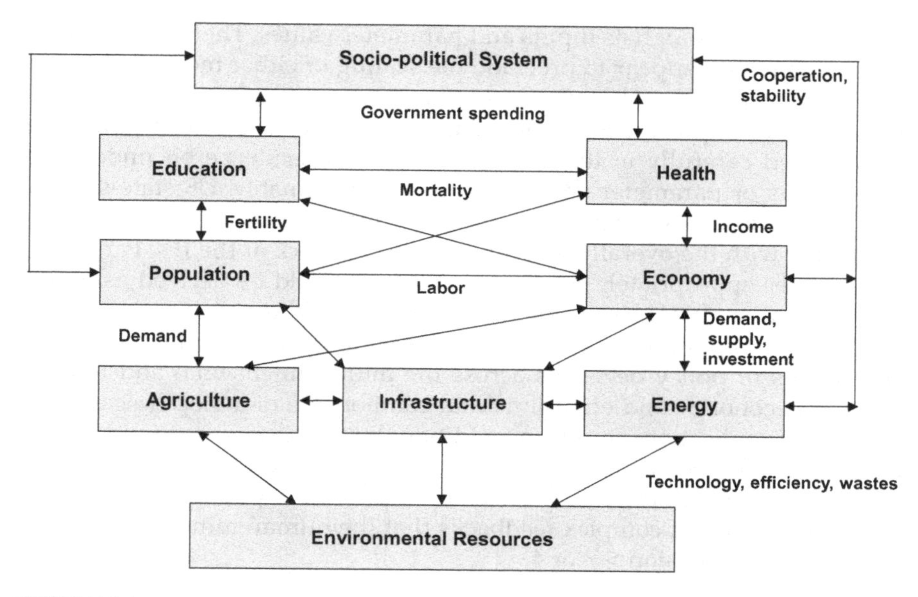

FIGURE 10.1
Schematic illustration of the international futures (IFs) model (adapted from Hughes 2009).

Land and water use, in combination with technological efficiencies and population, impact agriculture. Infrastructure is central to population, economy, agriculture, and energy. In turn, energy influences infrastructure, the economy, and environmental quality (Figure 10.1).

The mathematical formulations that underlie the IFs are grounded in theory (Hughes 2009) and supported by substantial data. The global domain represented by the IFs is built spatially using country-specific data developed for nearly 100 countries. This spatially explicit modeling approach differs fundamentally from homogeneous representation of global dynamics as described by the World3 model used by Meadows et al. (1972) to generate the simulations that formed the basis for the *Limits to Growth*.

Hughes (2019) emphasizes that the IFs model was developed mainly as an educational tool and was not intended to predict the outcomes of modeled scenarios. Nevertheless, the IFs model provides a compelling mathematical and empirical framework for exploring the possible implications of policy and decision making in relation to sustainability. From that perspective, the IFs model might well inform the design and evaluation of alternative suggestions for the form and function of an economy based on biophysical principles.

10.6.1 Scale Sufficiency

The IFs model is global in extent because of the geographic sum total of its country-specific underlying data sets. The World3, as a point model, becomes global through its inputs and parameter values. There is nothing in theory that would appear to preclude the scaling of either model to confine its spatial extent to the United States – the principal focal geography of this book. Individual process-level formulations in either model would have to be examined carefully in terms of any spatial dependence on underlying assumptions or parameter values and units. Presumably, US state-specific data sets could be developed to replace country-specific data and remain compatible with the overall computational framework of the IFs. Parameter values scales appropriately to the United States could be derived as necessary for a national-scale implementation of the World3 model.

Nation-scale applications of these models would permit exploration of the implications of policy decisions across the multi-dimensional and interrelated socioeconomic and environmental components of a biophysical economy. Model applications could lend additional strong inference in suggesting specific outcomes anticipated for policy decisions in question. The models might valuably point out unexpected consequences of policy decisions that emerge as a result of complex feedbacks that the human mind cannot track in systems of dimension > 3 or 4.

10.7 Green Economy Policies

Policy can be viewed as defining the will of people and organizations to effect change. Demonstrating the will to design and implement a biophysical economy is as important, perhaps more important, as having the technological capabilities to enact the requisite economic transformation (Costanza 2010).

In this vein, green economy policies can be defined to include green fiscal policy, green investment, and green jobs aimed at sustainable economic development (Wang et al. 2022). Green fiscal policy can manage collecting and spending money through government regulation to achieve desired social and economic objectives. Green economy policies can be used to influence investments towards those that are environmentally sound and consistent with sustainable economic development. Green policies, including environmental subsidies, carbon capture and storage, emission trading and pollution discharge fees, can reduce emissions, conserve energy and resources, and create green jobs (Wang et al. 2022). These authors discuss the importance of green policies in directly contributing to sustainable economic development (Figure 10.2). Just as important, green policies can promote entrepreneurship that fosters innovations in green technology – or ecopreneurship. The green technological advances can then contribute to overall sustainable economic development. Potential (positive) feedback from a sustainable economy on further ecopreneurship can accelerate the development

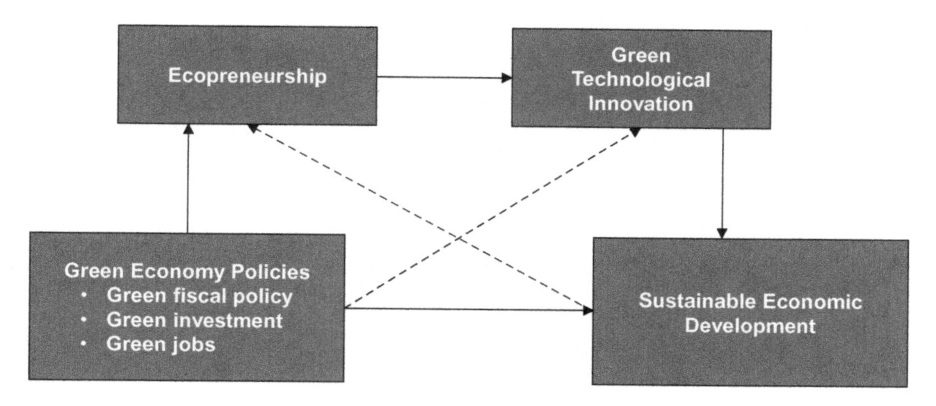

FIGURE 10.2
Conceptual framework that relates green economy policies to directly and indirectly to sustainable economic development (from Wang et al. 2022).

of a sustainable economy. Similarly, green investment and green jobs can lead directly to green technological innovation in their green policies framework (Figure 10.2).

An important objective of the presentation of the green policies framework was to emphasize the role that such policies play in promoting environmentally sound (green) business models and advances in technology that work towards sustainable economic development, including a biophysical economy.

Section 3

Synthesis

11

Feedback Mechanisms

11.1 Introduction

Building and operating a sustainable economy depends importantly on the recognition of feedback mechanisms that characterize complex socioeconomic and ecological systems (Giampietro and Mayumi 2018). Feedback is established when the results of an action serve as input information to another part of the system, which establishes a "cause and effect" relationship among the interconnected components and generates a loop structure in a conceptual model of the system. The existence and number of multiple, linear, and non-linear feedback mechanisms – some with thresholds and time delays – largely determine the complexity and dynamic behavior of the system (Sterman 2000; Allen and Starr 1982). In developing a biophysical economy, some feedback mechanisms that operate independently of the underlying economic paradigm are already known and need merely to be accurately accounted and correctly incorporated into the economic transition and its associated conceptual model. For example, feedback relations between supply and demand will likely remain important regardless of the enacted economic paradigm.

Designing a biophysical economy will additionally require recognition of feedback mechanisms that are not explicit in the neoclassical model (Hall and Klitgaard 2018). Relationships between utilization of renewable resources and their rates of replenishment have always been in effect, regardless of the economic system, but simply not directly accounted for in considering sustainability. Undoubtedly, feedback relationships among key socioeconomic and ecological system components exist that remain to be discovered. One intention of forcing the conversation by way of this book is to stimulate investigations that might identify new feedback mechanisms important to sustaining an economy, but mechanisms that remain overlooked or ignored in description of or operating within the prevailing neoclassical economic paradigm.

DOI: 10.1201/9781003308416-14

11.2 Positive Feedback

A positive feedback occurs when, for example, the product of a reaction leads subsequently to an increase in that reaction. If the net gain in the loop is positive and >1, an exponential increase is commonly observed. Recall the mathematics of "greater than exponential" or supernormal human population growth and its implications for infinite resources demands in real time – the finite singularity (West 2017). In complex dynamic systems, positive feedback loops tend to force the system away from any quasi-equilibrium. They contribute in that sense to system instability. Singularity – infinite or finite – appears inarguably destabilizing for the human enterprise and ecosystems in general on a finite planet.

Examples of positive feedback for phenomena relevant to global life-support systems are readily found in the study of climate change. Increased temperatures increase the melting of snow and ice at high latitudes and altitudes that reduce albedo and lead to further increases in temperature and further increases in melting. Increases in solar radiation associated with reduced cloud cover increase drought conditions, which, in turn, further reduces evaporation and cloud cover, which further dries the landscape. Increases in CO_2 emissions lead to increases in air temperature and humidity that correspondingly increase the use of air conditioning, which leads to additional increases in CO_2 emissions, which eventually requires more air conditioning – and the feedback continues until no more air conditioning is possible.

The demand for continuous growth in neoclassical economics can be interpreted as a positive feedback loop, where profit derived from growth is used to create conditions to grow further and accrue additional profit, some of which is used to continue the positive loop dynamics. The underlying assumption of the neoclassical paradigm is that this loop can continue to cycle indefinitely. Any increasingly scarce resource that might constrain growth (negative feedback) will be readily supplanted by some replacement as the result of continued technological advances afforded, in part, by investments in technology enabled by profits.

The design and implementation of a biophysical economy will benefit from the identification of sources of positive feedback in the transformational and evolving managed socioecological system that will be required. Comprehensive, integrated socioecological systems models can contribute to the identification and isolation of destabilizing positive feedback mechanisms and further the development of a sustainable economy at scale.

11.3 Negative Feedback

Oxford defines negative feedback as a counteraction of an effect by its influence on the process giving rise to it. In terms of homeostasis, negative feedback counteracts change and returns a diverging parameter back towards its original set point. Therefore, negative feedback tends to stabilize dynamic systems. Density-dependent growth rate is a common example, and one certainly relevant to the design and establishment of a biophysical economy. Initially high growth rates become reduced as the overall magnitude of the growing item (e.g., population) approaches some upper limit (e.g., carrying capacity) as necessary sustaining resources (e.g., food, water) become increasingly scarce. In social parlance, negative feedback can simply take the form of criticism in response to some verbal or behavioral activity.

Similarly, the design and construction of a biophysical economy can be facilitated through the identification of key negative feedback mechanisms in an evolving socioecological system. The integrated models can likewise help in the identification and, if necessary, amplification of negative feedback that lends stability in overall system performance.

11.4 Properties of Dynamic Systems

The structure and interconnectedness of the system determine its dynamics. Understanding a dynamic system distills to an identification of the source and nature of positive and negative feedback loops that exist among interrelated variables that define the system. As system complexity increases, the challenges of accurately describing the underlying loop structure and overall dynamics increase correspondingly.

Interestingly, dynamic systems are rather limited in their overall behavior, but their behavior can help an observer in identifying the underlying structure and feedback of the system (Figure 11.1). The dynamics illustrated in this figure are for one-dimensional systems, but the descriptions extend to any arbitrary number of dimensions necessary to describe complex socioeconomic and ecological systems. Figure 11.1(a) shows simple nonlinear (exponential) growth in relation to a similarly ever-increasing carrying capacity. This is essentially the neoclassical infinite world model. Endless growth implies limitless prosperity and, thus, this model seems to define the best of all socioeconomic, but perhaps not ecological, worlds. The model works as long as it works. Note that each of the three remaining system behaviors all demonstrate some initial non-linear growth phase until limitations feed back to constrain further growth. The public and private sector institutions that remain committed to the current neoclassical economic model will strive to maintain limitless growth at all costs. West (2017) reminds us that one of the inescapable properties of exponential growth is the singularity, which offers some comfort in requiring infinite time to manifest. However, for systems growing faster than exponential (e.g., human population), the singularity – where demand for at least one system requirement moves to

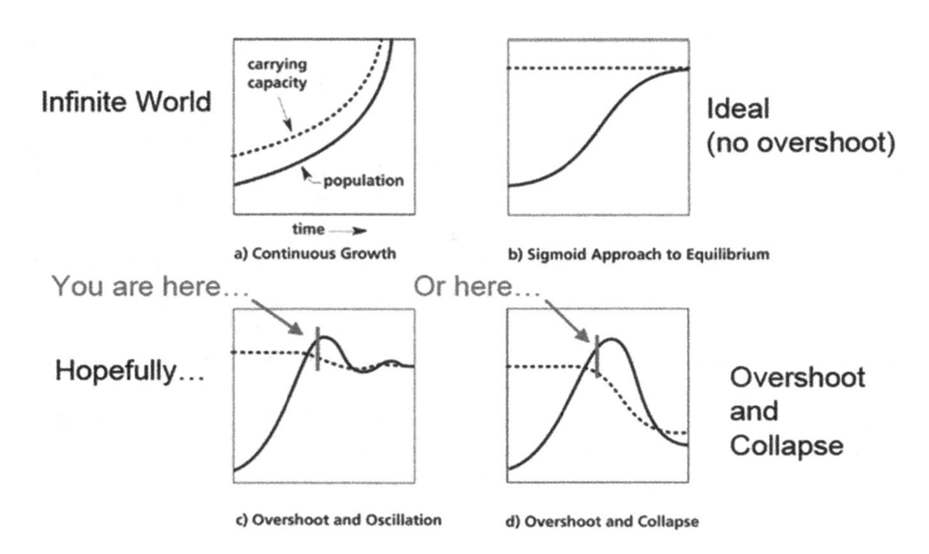

FIGURE 11.1
Possible characteristic behavior of dynamic systems (from E. Kalnay, personal communication).

infinity – occurs in finite time. This does not portend well for superlinear systems on a finite planet.

A second form of dynamic system behavior shows an initially slow growing population that subsequently accelerates in growth and then decelerates towards some constant and finite carrying capacity (Figure 11.1(b)). This behavior implies immediate and completely accurate feedback between the growing population and conditions that define the carrying capacity. This dynamic describes a perfect feedback and management system – certainly a desirable circumstance for adaptively managing a complex socioeconomic and ecological system. A constant carrying capacity is accurately known. The management pathway is perfectly understood and there are no time delays or thresholds that add uncertainty to the management process. The system is entirely determined and the desired outcome assured. This dynamic condition describes the best of all worlds in the context of sustainability. This condition likely does not exist in the measurable world – with perhaps the exception of bacteria growing in a laboratory culture.

If the carrying capacity is not constant, but fluctuates about some positive value, and feedback includes some time-delay, the growing population can overshoot the carrying capacity and then oscillate around the changing carrying capacity (Figure 11.1(c)). The population essentially attempts to track a changing carrying capacity in a system defined by dynamic constraints on population size. This behavior underlies the conceptual model for realistically addressing sustainability in the context of a biophysical economy. The abundance and availability of potentially limiting, but manageable, renewable resources will fluctuate in space and time – in part as the result of management and partly in response to factors not under the control of decision makers. The results of managing adaptively given imperfect understanding and limited measurements concerning possible limiting resources-establishes the situation where the system attempts to track a dynamic carrying capacity. As long as management is sufficiently successful in not depleting the renewable resources, the system might persist indefinitely – that is, be sustainable.

If the carrying capacity is declining through time, the population can overshoot and collapse (Figure 11.1(d)). The continued exploitation of an eventually limiting resource through mismanagement or total absence of management can lead to the collapse illustrated in Figure 11.1-(d). This is essentially the dynamics displayed by the World3 model and the "business as usual" scenario described by Meadows et al. (1972), *Limits to Growth*.

An important point is that despite the implied (and unknown) complexity of a nation-scale biophysical economy, its functional dynamics will be consistent with one of the patterns in Figure 11.1, although the exact pattern might change as the economy evolves. However, a robust monitoring scheme can provide data that can help identify the particular dynamic in operation with sufficient lead time to make necessary adjustments in the design and execution of the dynamic biophysical economy.

11.5 Dynamic Carrying Capacity

The concept of dynamic carrying capacity appears central to the design and implementation of a biophysical economy. The distribution and abundance of naturally renewable resources (e.g., fisheries, forests, wetlands) vary in space and time. Fish populations are notorious for highly variable populations sizes in relation to ecological and environmental conditions that influence population dynamics. A single strong year class defined by unusually high survival of early life stages can generate longer-term abundance in the face of subsequent years characterized by typical early life stage mortality.

A key objective in managing renewable resources to sustain an economy based on biophysics is to manage carrying capacity where possible and manage towards carrying capacity when carrying capacity is not manageable. Humans continue to demonstrate an increasing ability to manage fisheries, forests, and modern agriculture with some success in managing water resources. Taken to the extreme of highly managed aquaculture, fishery managers have become arguably proficient at managing habitat, food supplies, disease, and environmental conditions to maximize sustainable yields. Comparatively recent management approaches have emphasized ecosystem management instead of focusing on the small number of populations of commercial interest. The objective is to maintain carrying capacity within some

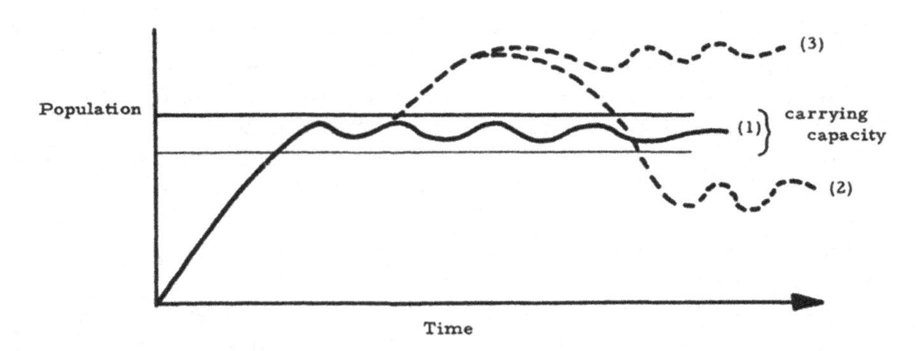

FIGURE 11.2
Concept of dynamic carrying capacity (after USEPA 1974).

predictable range that is consistent with obtaining managed resources (e.g., fish, timber) at some corresponding range of production. This is illustrated by case 1 in Figure 11.2.

Alternatively, management might focus on increasing carrying capacity as in case 3 (Figure 11.2). Modern agriculture is an example of increasing energy availability (fossil fuels), fertilizers, pesticides, and increased knowledge have resulted in increased carrying capacity defined by corresponding increased yields per ha.

The carrying capacity of renewable resources can be pushed past a point where capacity diminishes (Figure 11.2, case 2). As previously described, populations defined by a diminishing carrying capacity tend to overshoot and collapse.

11.6 Ecosystem Management and Feedback Mechanisms

Implementing a biophysical economy will require holistic management of complex, adaptive socioecological systems that comprise hierarchical organizations of structures and processes that manifest as positive and negative feedback mechanisms (Giampierto and Mayumi 2018; Allen and Starr 1982). Ecosystem management has been developed to integrate scientific understanding of ecological interrelationships and socioeconomic and political values systems directed towards the long-term sustainability of ecosystems (Pavlikakis and Tsihrintzis 2000; Christensen et al. 1996; Grumbine 1994).

Grumbine (1994) offered ten major aspects of ecosystem management as

1. Hierarchical context
2. Ecological boundaries
3. Ecological integrity

4. Data collection
5. Monitoring
6. Adaptive management
7. Interagency cooperation
8. Organizational change
9. Humans embedded in nature
10. Value systems

Ecosystem management differs from more traditional resource management by focusing on the longer-term sustainability of the entire ecosystem in contrast to shorter-term yields and economic gains from selected resources (e.g., fish populations, forests). Correspondingly, ecosystem management emphasizes protection and or preservation of biodiversity across all relevant scales instead of a single species. Finally, ecosystem management recognizes that humans are components of the ecosystem and human uses need to be considered in relation to overall management of the system of interest.

Ecosystem management, based on sound ecological theory and principles, can in theory be applied to any scale. Pavlikakis and Tsihrintzis (2000) offer the following generally recognized principles of ecosystem management:

> A collaborative approach must be developed among participating landowners, scientists, regulatory agencies, and other stakeholders to define the boundaries of the system of interest, set goals and objectives, and address any conflicting legal mandates.
>
> The sensitivities and cultural values of regional inhabitants (e.g., habits, traditions, religion, etc.) must be taken into account when developing and implementing ecosystem management. Human perceptions, well-being, and aspirations must be addressed in striving for long-term sustainability.
>
> Ecosystem management recognizes the importance of multiple human uses and activities at local scales, while also protecting the fundamental integrity of the managed ecosystem. Ecosystem management respects the rights of private ownership while obliging restrictions where necessary for the larger public interest.
>
> Ecosystem management should be based on the best available science and data developed by a multidisciplinary team focused on the overall planning, design, and decision-making process. The decision-making process should be developed recognizing inherent uncertainty and lack of complete understanding of ecosystem structure and function.

Successfully applying these principles across multiple scales (local, regional, national) of ecological, environmental, social, and political systems, characterized by complex feedbacks, will be required in the design and implementation of a sustainable, biophysical economy. Pavlikakis and Tsihrintzis

(2000) outline an eight-step methodology for ecosystem management aimed at a principled approach (Table 11.1).

Christensen et al. (1996) emphasize the need for long-term commitment in planning and implementation of ecosystem management. Successful application of ecosystem management also requires institutions that are adaptable as ecosystems and human needs and interests change. Ecosystem management will benefit as more professionals demonstrate understanding of scientific and socioeconomic issues and exhibit the ability to communicate these

TABLE 11.1

A Principled Methodology for Implementing Ecosystem Management (Pavlikakis and Tsihrintzis 2000)

Step	Description
Identify local issues	Establish the key issues to be addressed; identify local needs, areas of conflict and collaboration; identify opportunities, expectations, limits, and restrictions brought by participants
Participation	Identify stakeholders and encourage participation; provide necessary information and training; enlist support from the mass media, local government and private organizations, NGOs
Political, legislative, and economic analysis	Garner support from local politicians and activists; identify and understand relevant legislation; understand mission and operations of federal, state, and local agencies; describe the economic conditions of the region potentially impacted by ecosystem management
Definition of goals	Specify the goals and objectives, including criteria for success/failure; describe the desired future trajectory and behavior of the successfully managed system; accommodate human use and occupancy
Delineate ecosystem boundaries	Ecosystems are open and dynamic, but operational boundaries will be needed to define the system subject to management; identify ecological, socioeconomic, and legislative or regulatory boundaries of the overall management application
Development of a plan	Develop a holistic approach for implementing ecosystem management for the application; solicit input from scientists, engineers, and stakeholders, including government, local public and private participants (landowners, businesses, agencies, private organizations); develop decision-support systems to facilitate management and decision making within an adaptive framework
Monitoring	Construct a statistically defensible program for obtaining data and information that describes the pre-selected ecological, environmental, and socioeconomic responses of the system to management actions; quantify spatial-temporal variability and uncertainty in system responses to management actions to inform robust sampling programs in support of monitoring
Evaluation	Examine the results of the monitoring program in relation to the specific metrics used to judge the effectiveness of management in achieving the stated goals and objectives; revise other steps in the methodology as necessary within an adaptive management framework and approach to decision making

diverse issues with scientists, managers, and stakeholders. Importantly, ecosystem management recognizes the importance of human needs but at the same time underscores the fact that ecosystems have limited capacity to meet ever-expanding demands while maintaining their structural and functional integrity (Christensen et al. 1996).

11.7 Feedback, Carrying Capacity, and a Biophysical Economy

A biophysical economy will comprise a complex set of socioecological structures and processes that will be hierarchical in structure and dynamic in behavior. This economy will evidence sophisticated arrangements of disparately scaled structures and interrelated feedback mechanisms – both positive and negative. Understanding how to identify, as well as create and manage, feedback in the system in a way that confers longer-term stability will be a necessary ingredient in designing a biophysical economy.

Despite the challenges implied by the complex socioecological system that will manifest as a biophysical economy, its characteristic overall dynamics will be knowable. The challenge will be to manage the economy to some dynamic carrying capacity that will be partly under human control, but also impacted by the natural and/or managed rates of renewal of key limiting resources. Importantly, carrying capacity will undoubtedly vary by region and be defined by different potentially limiting resources in a nation-scale biophysical economy. The carrying capacity will be dynamic, and management institutions will require the flexibility to adapt to changing resource availability and potential limitations as they impact regional and overall economic performance and long-term viability (sustainability).

Ecosystem management offers a comprehensive and adaptable framework that can support the challenges in designing and operating a biophysical economy. The eight-step process outlined in Table 11.1 can be developed and put in place at local, regional, and national scales. The process can be informed through the corresponding collation and analysis of data currently collected by private and public agencies and organizations as described in previous chapters. Regional economic models can also be constructed to facilitate data analysis and integration in support of managing the complex socioeconomic and ecological systems that together make up a nation-scale biophysical economy. The ecosystem management process appears well suited to address the complex feedback mechanisms implied in the integration and management of linked social, economic, and ecological systems – each with their own characteristic hierarchical, dynamic, and evolving structure and function.

12

Models and Economic Transformation

12.1 Relevance and Contribution of Models

Mathematical causal representations of socioeconomic systems will contribute to the design and implementation of an economic model grounded in biophysics. Simple extrapolation of historical data might prove insufficient to project future economic conditions from current neoclassical assumptions when the intent is to fundamentally change the rules.

Mechanistic or process-level models can facilitate the change in economic course in at least two ways. Applicable models can be used to forecast future economic conditions and the sustainability of supporting biophysical subsystems in the face of change in the underlying socioeconomic paradigm.

Analysis of the structure and function of the models can further the understanding of how the interactions among model components and processes produce the economic forecasts. Concepts and methods of quantitative sensitivity and uncertainty analysis can be used to characterize the impacts of variability and uncertainty on model results. These approaches readily identify the dependence of specific model results on model structure, input data, and model parameter values. The results of sensitivity and uncertainty analysis provide useful information concerning the precision of model forecasts of future economic conditions associated with a transition to a biophysics-based economy. These kinds of analyses also pinpoint the key modeled structures (e.g., state variables) and processes that determine specific model results. This information can usefully guide improvements to the model and identify critical data needed to improve model performance.

Models will likely be needed and implemented at various scales and complexities to help guide the transition from the current neoclassical economy to a biophysical economy. One can easily envision at least three scales of model applications that will be used to help build a biophysical economy: global, local or regional, and corporate models.

DOI: 10.1201/9781003308416-15

12.2 Global-Scale Models

Sustainable development ultimately has meaning only when implemented at a global scale. The economic costs of unsustainable practices operational anywhere on the planet ("leakage") will propagate to larger-scale systems required to assimilate the leakage. Global-scale models will be necessary to help understand and project the propagation of unsustainable activities, perhaps ongoing at multiple locations. These capabilities at the global scale can strategically assist with the design of a biophysics-based economy that minimizes leakage and informs corresponding socioeconomic activities and policy aimed at assimilating leakage and restoring overall global sustainability. Think globally, act globally.

To remain comprehensible and tractable, global scale models necessarily aggregate finer-scale structures and processes. For example, the World3 model is essentially a point model – absent any detailed spatial representation. Its mathematical formulations become global in the scale of application through the derivation of the input data that drive the model and the values of parameters that determine the inner workings or dynamics of the model. The World3 describes a homogeneous planet and focuses mainly on the longer-term global implications of interconnected economic, environmental, and demographic subsystems in relation to sustainability outputs that are correspondingly highly aggregated and globally scaled.

There are less aggregated global models of integrated socioeconomic and environmental systems. The International Futures (IFs) model achieves global scale by including data for individual countries (Hughes 2019, 2009). Figure 12.1 illustrates the diverse socioeconomic, political, and environmental components and process-level interconnections that define the IFs model (2009).

12.3 System of Environmental-Economic Accounting (SEEA)

The SEEA derives from the System of National Accounts (SNA 2009). The SNA comprises a statistical framework that includes macroeconomic accounts to inform policymaking, analysis, and research at a national scale. The SNA is produced under the auspices of the United Nations, the European Commission, the Organisation for Economic Co-operation and Development (OECD), the International Monetary Fund, and the World Bank (SNA 2009). The SNA summarizes the exchange of goods, services, and assets among non-financial institutions, financial institutions, government, NPI serving households, and households. These five sectors define the total economy, and valuation in the SNA is purely transactional. The primary accounts that

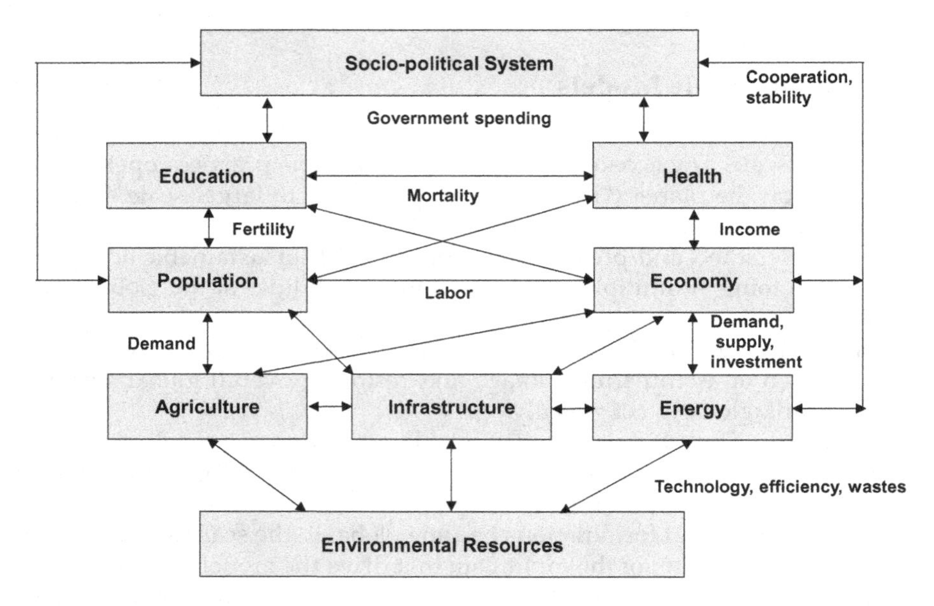

FIGURE 12.1
Schematic illustration of components and functional linkages in the IFs model (from Hughes 2009).

define a national economy in the SNA (and the SEEA) include production, distribution of income, capital, and finance (Table 12.1).

The rules and procedures underlying the SNA derive fundamentally from traditional business accounting. The double-entry bookkeeping practice is the basic unit in national accounting. The SNA was designed for flexible implementation using various accounts, classifications, and sectors. Data describing accounts can also be rearranged in the form of a social accounting matrix (SAM) to facilitate analysis and inform policy (SNA 2009). However, the intent here is not an in-depth description of SNA; the 700+ page SNA (2009) documents the SNA in excruciating detail. The key point is that the national accounts produce the gross domestic product (GDP) used to summarize economic activity. The GDP can be roughly defined as the monetary measure of the market value of the final goods and services produced by a country in a specific time (e.g., annually).

Importantly, from a biophysical perspective, GDP does not include depletion or degradation of environmental resources required to support production (Hall and Klitgaard 2018; Obst 2015). SEEA was developed to remedy, at least in part, this deficiency in reported values of GDP (Obst 2015). This framework measures the contribution of the environment to the economy and impacts of the economy on the environment and stocks of environmental resources (Banerjee et al. 2016). The SEEA incorporates accounting processes that include environmental data and information in physical units (e.g., energy use, water, and waste generation). Extending the computation

TABLE 12.1

Basic SEEA Sequence of Accounts

	Production Account
Main entries	Output, intermediate consumption, consumption of fixed capital, depletion
Balancing items/Aggregates	Gross value added, gross domestic product, depletion adjusted net value added, depletion adjusted gross domestic product
Distribution and use of income accounts	
Main entries	Compensation of employees, taxes, subsidies, interest, rent, final consumption expenditure, consumption of fixed capital, depletion
Balancing items/Aggregates	Depletion adjusted net operating surplus, depletion adjusted net national income, depletion adjusted net saving
Capital account	
Main entries	Acquisition and disposal of produced and non-produced assets
Balancing items/Aggregates	Net lending/borrowing
Financial account	
Main entries	Transactions in financial assets and liabilities
Balancing items/Aggregates	Net lending/borrowing

Source: United Nations et al. (2013) as cited in Obst (2015)

of GDP to include environmental aspects addressed by the SEEA affords the advantage of maintaining regular economic measures without having to offer alternative indices (HDI, GPI), which further requires justifying the use of new measures and gaining their broad acceptance within the communities of economic analysts and policymakers.

The SEEA provides for modifications of key accounts to more realistically incorporate the environmental impacts of an economy (Table 12.1). The components of production are adjusted to include the effects of resource depletion and provide for depletion of adjusted net value added and depletion adjusted GDP. Depletion is also considered in assessing the distribution and use of income. Depletion adjustments are made to net operating surplus, net national income, and net saving.

12.4 Regional Models

A nation-scale approach to economic sustainability might be informed through the application of regional modeling approaches. Beaussier et al.

(2019) reviewed a variety of regional economic modeling typologies ranging from input-output analysis to agent-based and system dynamic models (Table 12.2). The review described the overall structure and formulation of the modeled approaches with considerations of spatial and temporal scales in application along with recognized strengths and weaknesses. The implication of a regional modeling approach is that a sufficient number of regions throughout the United States could be modeled to produce a nation-scale capability to explore alternative designs and pathway for implementing a biophysical economy.

One challenge in building a nation-scale economic model as an interconnected set of regional models concerns the availability or region-specific data to support the modeling effort. Grosskurth (2007) additionally addressed differences in regional conceptual models developed by managers to describe sets of socioeconomic processes of importance with the capabilities of existing regional models to quantitatively characterize those processes. Reconciling such differences can help guide model selection, model development, and model application to inform regional socioeconomic planning and analysis.

12.5 Corporate Models

In the context of sustainability at the scale of individual corporations, an increasing number of corporations are enacting management programs that include activities under three broad categories of environment (E), social (S), and governance (G) – or ESG (Henisz et al. 2019). Corporate management guided by ESG analysis and reporting is replacing historical centralized command and control as an alternative approach to enhance corporate performance and contribute to corporate sustainability (Katsamakas et al. 2022; Tijani and Ahmadi 2022; Armstrong 2020; Cornell and Damodaran 2020; Monciardini 2012).

A growing and evolving body of technical literature continues to characterize empirical relationships between corporate ESG factors and corporate sustainability (Oprean-Stan et al. 2020; Henisz et al. 2019). Firms enact ESG management programs with the belief that competent ESG activities contribute towards corporate sustainability (e.g., Orlitzky et al. 2003; Margolis et al. 2009; Eccles et al. 2014; Kotsantonis et al. 2015). Sassen et al. (2016) reviewed the literature and concluded that firms with substantial programs in corporate social responsibility (CSR), measured by ESG factors, generally demonstrated negative relationships between CSR performance and firm risk, where risk is defined as the cost of capital.

At the aggregate level, there appears a general relationship between corporate performance and ESG programs. For example, Kumar et al. (2016)

TABLE 12.2

Main Characteristics of Regional Economic Modeling Tools

	Economic Models					
Characteristics	IO and SAM*	Econometric	General equilibrium	Partial equilibrium	Agent-based	System dynamics
Formalizations	Tables of linear relationships	Regression relationships	Supply and demand equilibrium functions	Supply and demand equilibrium functions	Behavioral rules	Stocks, flows, and feedback loops
Equations	Linear	Linear and non-linear	Linear and non-linear	Linear and non-linear	Non-linear	Non-linear
Timescale	Static	Dynamic	Static and dynamic	Static and dynamic	Static and dynamic	Dynamic
Geographic scale	Meso- to macro-	All scales	Macro-oriented	Meso- to macro-	Very macro- or micro-	Very macro- or micro-
Strengths	Tracks interindustry linkages; easy to implement	Accurate short-term forecasts	Endogenous prices and substitution effects	Endogenous prices and substitution effects; simpler than GE	Models agent behaviors and interactions	Includes relevant variables and complex interactions
Limitations	Fixed prices; no substitutions	Predictive power tied to data quality	Large data requirements; black-box	Limited to a few sectors; less detailed economic indicators	Lacks standardization; intractable; black-box	Lacks standardization; intractable; black-box

Source: Beaussier et al. 2019

**Input-Output and Social Accounting Matrix*

analyzed data (2014–2015) from 966 corporations across 12 industry sectors and found that firms with high scores for ESG activities consistently demonstrated decreased volatility in stock price compared with firms that performed less convincingly in ESG.

The results are less clear when examining the contributions of individual dimensions of ESG to corporate sustainability (Sassen et al. 2016). Positive relationships between corporations and local communities (S), as well as relationships with employees (G), appear negatively correlated with firm risk (Bouslah et al. 2013; Oikonomou et al. 2012). However, mixed results were obtained in analyzing corporate environmental (E) activities in relation to corporate performance (Bouslah et al. 2013). Similarly, Kim and Li (2021) statistically analyzed corporate MSCI ESG data and S&P Capital IQ-Compustat financial performance data for 4,708 firms representing a wide range of industries over the period of 1991–2013. Their results based on simple correlations and multivariate regression indicated a positive relationship between ESG factors on corporate profitability with corporate governance having the greatest impact. The social factors of ESG demonstrated the strongest relationship with firm credit rating. Curiously, the environmental component of the total ESG score indicated a negative relationship with credit rating. In contrast, Park and Jang (2021) found that environmental and governance factors were more important in influencing investment decisions in an analysis of South Korean corporate performance and ESG data.

In addition to empirical analysis, dynamic simulation models can be used to explore causal relationships between ESG and corporate sustainability (Bakshi and Fiksel 2003; Cosenz and Bivona 2021; Cosenz et al. 2020, 2019; Lozano 2015; Sterman 2000). The models attempt to mechanistically describe the structure and function of corporate systems within a broader business, social, and natural environment (e.g., Fig. 12.2). The models capture the fundamental business operations of the corporation, while including specific inputs and outputs that directly connect the performance of the modeled firm to dimensions of ESG. For example, the generalized model in Figure 12.2 identifies investors as key stakeholders with functional connections to corporate strategic resources. Importantly, model outputs include metrics for environmental impact (CO2 emissions) and social value (community development and well-being, public spending, employment rate). This functionality permits explorations of modeled corporate ESG activities that change the dynamics of internal corporate operations and their corresponding effects on stakeholder, social, and environmental values. Governance activities are implicit in the modeled internal key processes, which, in turn, influence corporate revenue streams, firm value proposition, and outputs related directly to ESG accounting and reporting.

The modeling process can be used to explore the implications of alternative corporate ESG activities on corresponding anticipated effects on ESG accounting metrics. The general corporate model structure and function (e.g., Fig. 12.2) can be modified and refined to represent generalized firms within a

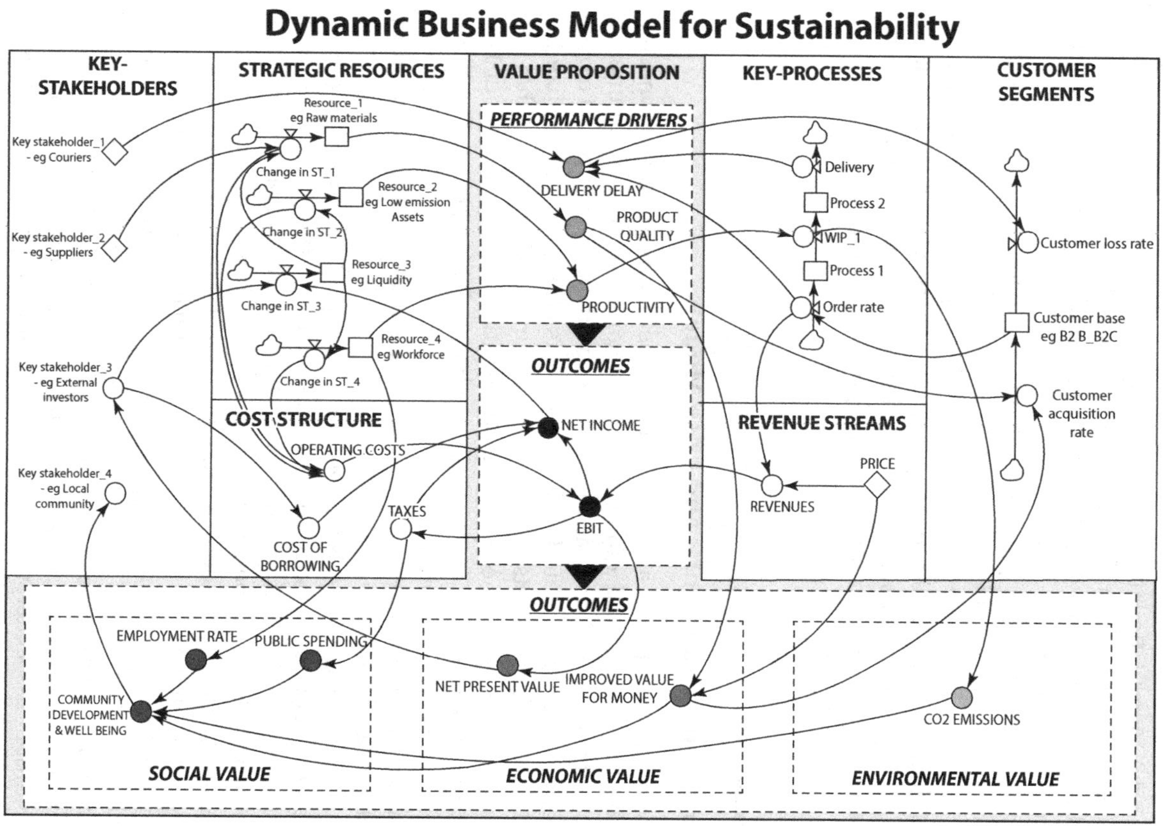

FIGURE 12.2
Schematic illustration of the conceptual model used to guide the development of the corporate sustainability assessment model, CorSAM V 1.00. (Taken from Cosenz 2019).

specific industry category (e.g., chemicals, utilities, health care, finance as in Sassen et al. 2016) or specific firms – both manufacturers and service providers (e.g., Cosenz and Bivona 2021).

Furthermore, the models can be implemented within a stochastic (e.g., Monte Carlo) computational framework to quantify the implications of variability and uncertainty in model inputs (e.g., reliability of raw material supply) and internal processes (e.g., labor force, productivity) on model results, including ESG metrics.

Importantly, the dynamic process modeling approach provides an opportunity to potentially validate and understand the previously described empirical relationships within a mechanistic context. Such causal understanding increases the ability of corporate managers and decision makers to enact ESG programs customized to optimize ESG returns and financial performance, and thereby contribute to corporate sustainability.

12.6 Environmental Profit and Loss Models

The environmental profit and loss (EP&L) methodology was pioneered by Kering to measure and monitor the monetary costs of environmental changes, both positive and negative, associated with the operations of a business (Kering 2013). Corporate EP&L assessments attempt to internalize the external natural capital costs of operations. Kering (2013) reported that collectively across the private sector these externalities amounted to about 11%, or $6.6 trillion, of global GDP in 2008, according to UN statistics. The EP&L approach is initially attractive because it presents its results in terms of monetary values (e.g., US dollars, euros, etc.) that are readily comprehended by corporate decision makers.

There are several motivations for undertaking corporate EP&L assessments. A firm might voluntarily perform an assessment to identify potential cost-saving, manage risk, or seize market opportunities. Firms that report EP&L assessment results might benefit from enhanced reputation and improved relationships with stakeholders. Regulations and legislation might enjoin a firm to report its impacts on natural capital through imposition of fines, taxes, or other penalties. Competition for increasingly scarce resources and associated changes in market structures and business models might stimulate EP&L analysis to help manage corporate operations in relation to increasing costs of raw materials and changes in demand for products. Changing operating conditions associated with increasing pressures on infrastructure, environmental degradation, more frequent extreme climate events, and dynamic social structures can disrupt corporate operations. EP&L can help develop better relationships with employees, customers, investors, and suppliers to enhance demand, increase revenue, and lower costs.

An EP&L analysis routinely consists of several steps supported by corporate-specific data and additional more generic data where necessary. Vodaphone (2014–2015) identified five steps in developing a corporate EP&L:

1. Delineating what should be included in the overall analysis. Generally, corporations have attempted to include as many aspects of their structures and operations in developing comprehensive EP&L assessments. For example, Kering, a purveyor of luxury items, included its stores, warehouses, transport, assembly, raw material processing, and raw material production in its EP&L analysis. This initial step also defines the kinds of environmental impacts that will be included in the assessment.
2. Mapping the value chain. The second step maps the value chain and can be informed by life cycle analysis from the supply chain. The mapping develops a clear picture of the interconnections of the processes from production of raw resources through product assembly and retail (Kering 2013).
3. Collect the supporting data. Primary data from the firm's own operations are of course critical to conducting the analysis. Secondary data from the supply chain and any additional research are necessary to extend the analysis beyond the immediate operations of the firm and develop a comprehensive assessment of the complete set of profits and losses relevant to the firm's overall operations.
4. Assign monetary values to the impacts, both positive and negative. This step is both a principal benefit and a challenge of the overall EP&L approach. Adopting EP&L offers the benefit of providing a common monetary denominator for profits and losses frames the analysis in terms readily understood by managers, regulators, investors, and stakeholders. At the same time, some members of the regulatory and stakeholder community decry the monetary valuation of environmental impacts as overly simplistic, fraught with inaccuracy, or simply immoral.
5. Complete the analysis and develop the EP&L report.

Other EP&L assessments define more steps (e.g., Kering 2013), but these additional steps are essentially disaggregation of the above five. For example, step 3 has been broken into additional steps that include a general step outlining the data collection process with additional steps that refer separately to collection of primary data and collection of secondary data. However, the overall approaches remain similar when viewed in total.

Kering (2013) identified six main topic areas of potential impacts that were included in their EP&L, including air pollution, GHG emissions, land-use changes, waste generation, water consumption, and water quality (Table 12.3). The study then addressed changes in environmental quality and human

TABLE 12.3

EP&L Measures and Values of Impacts (after Kering 2013)

Impact	Emissions, Resource Use (Units)	Environmental Change	Change in Well-Being
Air pollution	Pollutants, e.g., NOx, SOx, VOCs (kg)	Increased concentrations	Respiratory disease, agriculture losses, reduced visibility
GHG emissions	CO_2, N_2O, CH_4, CFCs (kg)	Climate change	Health impacts, economic losses
Land use	Tropical and temperate forest, wetlands (ha)	Reduced ecosystem services	Economic losses, reduced recreation, and cultural benefits
Waste	Hazardous and non-hazardous (kg)	Contamination	Reduced enjoyment of local environments, decontamination costs
Water use	Water consumption (m^3)	Increased water scarcity	Malnutrition and disease
Water pollution	Heany metals, nutrients, toxic compounds (kg)	Reduced water quality	Health impacts, eutrophication, economic losses

TABLE 12.4

Summary of Kering EP&L Assessment across Tiers (Adapted from Kering 2013)

Tiers	0	1	2	3	4	Total
$USD (millions)*	59.8	106.8	35.9	211.1	412.0	825.6
Percent	7	13	4	26	50	100

Source: Kering 2013, figure 9
* *Converted from euros, 1euro = $1.07*

well-being associated with each topic area (Table 12.4). Though developed independently, the impacts assessed by Kering map, in part, to the categories of planetary boundaries described in Chapter 3, including climate change, atmospheric ozone depletion, atmospheric aerosol loading, biochemical flows, freshwater use, and land-system change (Steffan et al. 2015).

With the intention comprehensively of addressing key processes and business components, Kering (2013) defined "tiers" within their supply chain for subsequent analysis:

- Tier 0 – stores, warehouses, and offices. This tier included direct operations and retail sales of Kering products.
- Tier 1 – assembly. Final manufacture or assembly of finished products (e.g., shoes, handbags, clothing).
- Tier 2 – manufacturing. Manufacturing of the sub-components used in the assembly of the final products for retail.

TABLE 12.5

Summary of Kering EP&L Assessment across Impacts (Adapted from Kering 2013)

Impact	Air Pollution	GHG Emissions	Land-use Change	Waste Generation	Water Consumption	Water Pollution	Total
$USD (millions)*	68.9	290.7	224.2	39.5	88.9	113.4	825.6
Percent	8	35	27	5	11	14	100

Source: Kering 2013, figure 9
* *Converted from euros, 1 euro = $1.07*

- Tier 3 – raw material processing. Transforming raw materials to subsequent materials (e.g., yarns, metals, leather) used in manufacturing.
- Tier 4 – raw material production. Primary production of raw materials used in processing (e.g., farming, mining, extraction).

Kering (2013) summarized the results of its EP&L across its operating tiers (Table 12.4) and categories of impacts (Table 12.5). Most of its operating externality costs were associated with Tier 6 raw material production (50%) and Tier 5 raw material processing (26%). Assembly (Tier 1) accounted for about 13% of total costs, while Tier 0 retail/offices and Tier 2 assembly contributed 7% and 4%, respectively (Table 12.4).

In comparison, Puma (2017), a major retail manufacturer of sporting shoes and apparel, estimated its tier-level percentage impacts as Tier 0 (5.3%), Tier 1 (24.9%), Tier 2 (4.5%), Tier 3 (29%), and Tier (4) (36.3%).

Summarizing Kering's EP&L results across impact categories showed that most of it impacts were in GHG emissions (35%) and land-use change (27%), while water pollution and water consumption accounted for 14% and 11%, respectively (Table 12.5). Air pollution impacts and waste generation rounded out the assessment at 8% and 5%. Puma (2017) produced similar percentage results for its impacts with GHG emissions (36.6%) and land-use (24.4%) dominating, while water use accounted for 16.5% of its 2017 impacts. Air pollution (8.3%), water pollution (7.7%), and waste generation (6.5%) contributed proportionally less to the overall EP&L results (Puma 2017).

The overall results appear sensible given that Kering basically transforms raw materials into high-end shoes, handbags, and apparel. The assessed impacts of raw material production and processing on GHG emissions and land-use impacts seem realistic for this firm. Importantly, the results of EP&L assessments will vary according to the operations and products (or services) provided by the firms adopting this approach. For example, Vodafone, a large mobile communications firm in the Netherlands, implemented an EP&L for 2014–2015 (Vodafone 2014–2015). Its estimated negative impacts total 21.6 million euros ($23.1 million), which were attributed to carbon emissions (42%), water pollution (33%), air pollution (24%), and waste generation

(1%). Vodafone estimated its positive impacts (benefits) as 37.4 million euros ($39.9 million) resulting from the estimated reduction of its customers' carbon footprint.

Conceptually, no barriers appear that would preclude implementation of EP&L at organizational scales in addition to individual businesses. In theory, EP&L could be applied to sectors within an economy, to regional economies, or to an entire national (global?) economy.

EP&L integrates concepts and methods fundamental to ecological economics with foundational aspects of a biophysical economy.

To honestly make use of the EP&L approach in supporting the design and operation of a BPE, the results of such assessments must be taken further. Regarding its initial assessment, Kering (2013) states:

> "The results are not related to Kering's financial results, past, present or future, and do not represent a financial liability or cost to Kering. ---- It is not intended to represent a forward looking statement or any financial obligation for Kering of any kind."

So, what is the point of an EP&L? It could be reasonably argued that if the assessed costs of operations are real enough to compute in the first place, which requires time and effort that cost real money to the firm and are of sufficient value for guiding management decisions aimed at reducing impacts (costs), then the resulting EP&L costs are real enough to be materially borne by the business entity. Otherwise stated, if the estimated values of corporate benefits are real enough to be legitimately used in offsetting impacts, the net monetary result is real enough that the business entity should reimburse society for net negative EP&L results and, in fairness, be reimbursed by society for net positive results. Note that the current neoclassical economic model allows for paying firms for their positive and negative impacts on environment and society in the form of subsidies and tax benefits. Precedent appears to exist. If it pays to grow, corporations should pay to grow.

13

Prescriptions for a Biophysical Economy

13.1 Introduction

The foundation has been laid (Richardson et al. 2023; Schwab 2021; Ji and Luo 2020; Sherwood et al. 2020; Klitgaard 2020; Hall and Klitgaard 2019, 2018; Yan et al. 2019; Moore 2015; Beinhocker 2012; Turner et al. 2011; Turner 2008; Bostrom 2009; Rockstrom et al. 2009; Costanza, Graumlich et al. 2007, Costanza, Leemans et al. 2007; Diamond 2005; Cohen 1997; Daly 1977; Meadows et al. 1992, 1972; Georgescu-Roegen 1971; Odum 1971; Lotka 1925; Malthus 1798; Quesnay 1765) upon which to build a biophysical economy (BPE). The following chapter prescribes an approach towards designing and building one.

The overall approach to realizing a BPE will first largely address the transition from the neoclassical firm and household model to one that more realistically describes economic activity in the real (physical) world. To that transition in the fundamental economic model, key concepts, observations, and methodological contributions from other topic areas that were described in previous chapters are considered accordingly. One topic, namely energy, which could have occupied a chapter on its own is introduced in relation to the sustainability of future economies, including the BPE. So the challenge of this chapter is largely one of integration of previously developed "pieces and parts" with the end game being an initial prescription for a biophysical economy.

13.2 A More Realistic Model

Central to the design of a BPE is a realistic model of an economy. The example selected for exposition here is taken from the Turner et al. (2011) description of the Australian economy (Figure 13.1). The stocks and flows model is similar in structure and content to those presented in Hall and Klitgaard (2018),

DOI: 10.1201/9781003308416-16

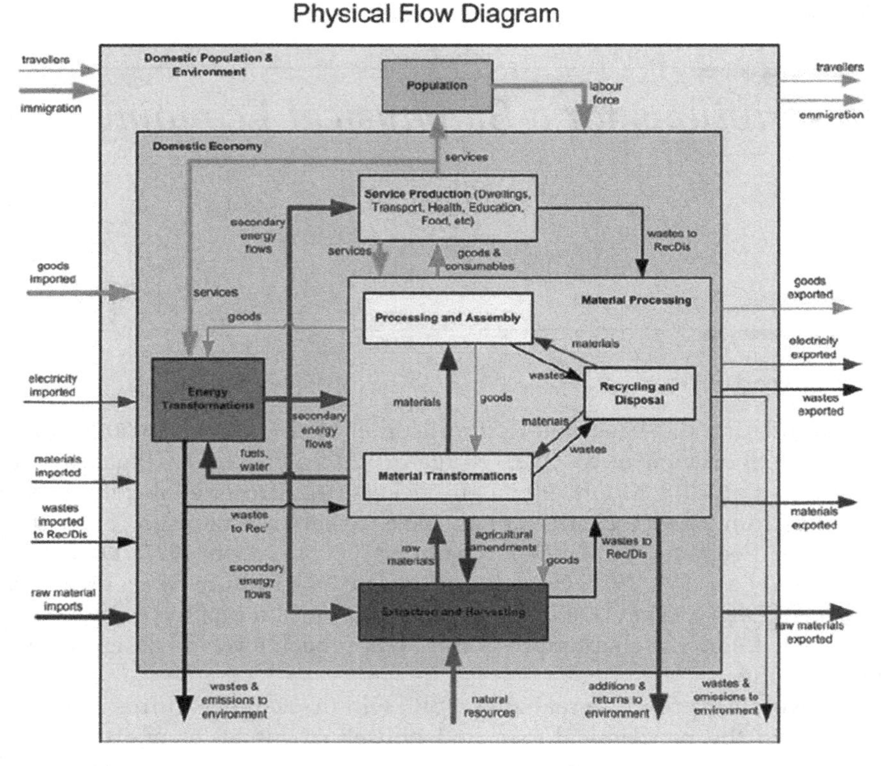

FIGURE 13.1
A schematic stocks and flows model of an economy adaptable to a biophysical-based economy (after Turner et al. 2011).

but with a focus on national scale, corresponding complexity, and associated detail. The model is offered in contrast to the neoclassical household-firm model (e.g., Chapter 1, Figure 1.1). The underlying assumption is that the basic structure of the Australian model applies generally to other nations, including the United States.

The Turner et al. (2011) model provides insight concerning the requirements of an operational BPE. The model importantly identifies the domestic economy as a subset of the overall environment and greater domestic population. For the sake of this presentation, domestic will equate to a national-scale BPE. A BPE is not separate from the environment, but congruent with it. Again, this structure underscores the dual nature of humans as simultaneously being "inside and outside" the natural system – breaking nation-scale constraints, but ultimately limited by finite global resources. Similarly, this structure recognizes that humans are also inside and outside the economy, as designated by a subset of the population serving as a labor force (Figure 13.1).

Within the greater domestic economy, the model identifies material processing, service production, energy transformations, and extraction and harvesting as important and interrelated structural components (Figure 13.1). Material processing involves the transformation of raw materials, which are, through processing and assembly, transformed into goods. Material goods and consumables, as indicated in the model, flow into the general economy within the context of service production. The model includes goods that are imported and exported, one of the functional connections of the national BPE to the global economy. The processing and assembly of goods also feeds back to energy transformations and extraction and harvesting. Material processing also includes the opportunity for recycling and the need for disposal. Recycling can provide materials used again in material transformation and processing and assembly.

Material processing depends on service inputs from the service production component and secondary energy flows from energy transformations. In turn, material processing can provide agricultural amendments (e.g., fertilizers, pesticides, machinery, technology) to the extraction and harvesting sector of the economy. Material processing can provide fuel and water used in the energy transformations sector. Material processing can also contribute additions or positive returns, as well as wastes, to the greater environment.

The service production component of the framework identifies dwellings, transportation, health, education, food, and other aspects of the overall economy not expressly related to material processing, energy transformation, or extraction and harvesting (Figure 13.1). The service sector benefits from goods and consumables produced by the material processing sector and, in turn, contributes to the production of those goods. The service sector requires energy and impacts that production of energy. This sector necessarily produces waste that can enter the recycling and disposal within the material processing sector. The model represents the overall impacts of the domestic economy on the greater population as feedback from services production.

The energy transformations sector provides the secondary energy flows that derive services production, material processing, and extraction and harvesting (Figure 13.1). Energy transformation, in turn, depends on inputs from services production (e.g., labor), as well as fuels, water, and goods (e.g., technology) from the material processing sector. In the Turner et al. model, the inputs of extracted fuels (e.g., fossil fuels) are provided through the material processing sector, which represents the costs of transforming raw to useable fuels. Energy transformation produces wastes that directly enter the environment (e.g., effluents, heat) or wastes that can be recycled or otherwise disposed (e.g., nuclear waste).

Extraction and harvesting provide the raw materials to the materials processing sector (Figure 13.1). Of course, extraction and harvesting depend upon the availability of natural resources as illustrated in the model. Extraction and harvesting depend importantly upon available energy to perform these functions and similarly require machinery and other technology. In this model, wastes associated with extraction and harvesting can enter the recycling component of the material processing sector, where some fraction can then enter back into the environment. The model does not show a direct connection to the services production center, but services at least in the form of labor are clearly necessary for extraction and harvesting.

Finally, the Australian stocks and flows model embeds the national economy within a global-scale consideration through pointing out the external inputs to the national economy of people (e.g., travelers, immigrants) as well as noting the importance of imported goods, energy (electricity), raw and other materials, and even wastes (Figure 13.1). Corresponding outputs or outflows from the Australian economy mirror the inputs in terms of people (travelers, emigration) and exported goods, electricity, processed and raw materials, and wastes (e.g., greenhouse gases).

The Australian model identifies structures and functions compatible with an operating BPE, but the model as illustrated in Figure 13.1 is not in and of itself a comprehensive design for a BPE. The following discussion suggests additions and modifications to the Australian model that could help transform it into a template for an economy more closely aligned with biophysics.

13.2.1 Desirable and Sustainable Stocks

The overall structure of the stocks and flows model is neutral concerning growth. The model could be used to emphasize continuous growth equally as well as sustainability. Using the model as a template to design a BPE requires the derivation of sustainable stocks throughout the socioeconomic system. The BPE is not necessarily anti-growth, but growth must be consistent with the available stocks required to support and sustain growth. Efficiency gains can, to a limit, permit and support growth while stocks remain constant. However, there are thermodynamic limits to efficiency gains and continued growth will, in the end, outstrip efficiency and leave a greater absolute gap to be somehow filled to sustain the system, i.e., Jevons paradox (Giampietro and Mayumi 2018; Alcott 2005).

13.2.2 Flows Compatible with Biophysical Constraints

Similarly, using the model to guide the design of a BPE depends upon the quantification of flows that are consistent with the derived values of sustainable stocks and planned growth. Measuring and managing flows is critical to the construction of a BPE given that desired sustainable stocks are simply the integration of inputs and outputs. Some flows can be measured directly; other flows can only be indirectly estimated from measured changes in stocks they connect.

Finally, the derivation of growth-sustaining stocks and flows must occur in concert. Both attributes necessarily integrate to define the overall dynamics of the system. An initial recommendation would be to develop a nation-scale version of the Australian stock and flow model for the United States. While perhaps daunting in complexity, derivation of such a model appears eminently feasible given rich sources of data that appear to be available from government entities, NGOs, and the private sector.

13.2.3 Population Size

Derivation of sustainable stocks and flows cannot be completed independent from considerations of overall population size and population growth. Resource requirements of a growing human population and indefinitely available resources that meet these requirements largely determine the size and potential for growth of a BPE. Implementing a BPE based on the model (Figure 13.1) will require the establishment of a sustainable population size. Setting efficiency gains in technology aside for the moment, a growing population will require a corresponding increase in resources, including energy. Clearly, to the extent that potentially limiting resources (peak oil, peak water, peak soils) can be increased through management, population size can increase. Therefore, to implement a BPE, the set of potentially limiting resources will have to be characterized at scales relevant to the population (i.e., local, regional, national, global). A non-negotiable limiting resource that has no substitute will ultimately define the sustainable population size, all else being equal.

Designing a BPE can be facilitated through the analysis of scenarios using the stock and flow model constructed for the United States for different population sizes under associated assumptions concerning energy and resource availability. Chapter 3 included an estimate of global population size of about 6 billion based on a simple allocation of energy produced globally based on a 100 GJ/y per capita requirement for an acceptable living standard. This simplistic approach could be applied using US energy production statistics to derive an initial ballpark estimate for a sustainable population in relation to reliable energy production. This estimate could serve as the basis for scenario development with subsequent modifications to include potential limiting factors in defining a sustainable population with an acceptable living standard.

13.3 Recognizing Boundaries

Planetary boundaries imply a global scale for evaluation (Rockstrom et al. 2009). However, in developing a nation-scale BPE, these same metrics can be applied at the corresponding scale. National data and information resources can be martialed to help determine the impacts of an operating economy on these planetary life-support systems (Table 13.1). The overall intent in creating a biophysical economy is to minimize impacts as much as possible and

TABLE 13.1

Proposed Planetary Boundaries (Adapted from Rockstrom et al. 2009)

Earth System Process	Data, Information Resources	Comments
Climate change	GHG emissions (Scopes 1, 2); generating utilities data	Quantify nation-scale emissions in relation to global boundary values
Ocean acidification	GHG emissions (Scopes 1, 2); generating utilities data	Same as above
Stratospheric ozone depletion	Any continued CFC (or equivalent chemical) use	<5% reduction from pre-industrial values
Atmospheric aerosol loading	Permits, direct measures	Quantify nation-scale emissions
Biogeochemical flows	Fertilizer use (purchases); soil runoff measures, models	P: 10-fold reduction N: about 25% of annual N fixation by national terrestrial ecosystems
Freshwater use	USGS, municipal utilities	<4,000 km^3 (4,000–6,000 km^3) – nation-scale portion
Land-system change	USDA NASS	≤15% of ice-free land surface (15%–20%) – national equivalent
Biodiversity loss	Habitat loss	Minimize habitat loss
Chemical pollution	NPDES permits, wastewater treatment	Minimize releases

keep economic activity within each boundary limit. Alternatively, the BPE can be evaluated to determine if a nation's contribution to potentially exceeding a boundary value is decidedly out of proportion to the impacts of other countries and the overall global community and accordingly enact policies to reduce impacts and strive to remain within the recommended boundaries.

A parallel activity might reasonably include the continued review and evaluation of the metrics used to quantify individual boundaries. In the spirit of adaptive management, it should be recognized that estimates of boundary values are inherently variable, scale-dependent, and uncertain. The use of planetary boundary values to guide the development and monitor the performance of a biophysical economy provides an opportunity for updating the boundary values in response to new information and increased quantitative understanding of the individual boundary phenomena and the impacts of economic activity on boundary metrics.

13.4 Adding Policy

As previously emphasized in Chapter 10, policy will play a key role in the design and particularly the implementation of a BPE and the stocks and flow diagram is absent of any explicit indications of policy considerations. As mentioned in Chapter 2, the World3 modeled scenarios that incorporated policy change, changes in cultural behavior, and technology did not result in economical simulated economic overshoot and collapse. With the right assumptions and corresponding model parameter values, the global simulations appeared indefinitely sustainable (Meadows et al. 1972). An initial attempt at the design of a sustainable economy based on biophysics might reasonably incorporate policy, behavioral change, and some aspects of technology (e.g., renewable energy, carbon capture and storage) to mimic the conditions that produced sustainable model results.

Policies that stimulate the transformation from an economy that emphasizes endless growth and consumption to a sustainable model that focuses on human well-being can facilitate the design and implementation of a biophysical economy (Costanza 2010). The green economy policies outlined in Chapter 10, including fiscal, investment, and employment activities, can provide a societal and cultural backdrop conducive to a biophysical economy (Wang et al. 2022). As further suggested in Chapter 10, policy considerations will likely dominate the technical aspects in deciding if and how to enact the necessary transformations to realize economic sustainability and set up a biophysical economy in the United States.

13.5 Adding Socioeconomic and Cultural Context

Different groups of people challenged by the same problem can require different solutions – an artifact of culture or a definition of culture (Little 2023). The diverse complement of richly different socioeconomic and ethnic groups that constitute society in the United States can stymie the adoption of something as novel as a new economic system, particularly if perceptions of unequal benefits manifest. Cultural divides and prejudices are not readily addressed in current conversations concerning the idea of economic transformation. One possible remedy to this cultural challenge lies in the transparent sharing of data and information used to execute an economic transition of nation-scale magnitude and importance. However, transparency also opens the door to misinformation, which in a social media world can quickly dominate the discourse and become entrenched in the belief and value systems of a sympathetic audience (witness the 2020 election lie).

At the same time, if disparate socioeconomic and cultural groups within the United States honestly realize and agree that a sustainable economy benefits everyone in the longer term, diverse people can galvanize a united response that can forcefully accelerate the adoption and implementation of a biophysical economy. Such consensus depends in no small part on increasing the general public understanding of structures and processes that determine the habitability and quality of life on Earth and define limits to cultural niche construction (Waring et al. 2023).

13.6 Financial Considerations

The establishment of a biophysical economy might both impact financial institutions and be impacted by finance (Chapter 9). The costs of conducting business as usual within the current neoclassical framework have been variously estimated as externalities. Damages to human health and infrastructure within the United States associated with pollution and other stressors (e.g., climate change) are real costs, and failing to address the underlying causal mechanisms will only continue and potentially increase these costs. A comprehensive accounting of the pollution costs is possible, as demonstrated by Muller et al. (2011). The accounting of gross external damages can be updated and used to evaluate the benefits of a biophysical economy in relation to future reductions in these costs. Such analysis could become a routine measure of the performance of a biophysical economy.

Chapter 9 also provided some estimates of the increasing costs of natural disasters which might be influenced (increased frequency, increased intensity) by climate change. Potential amelioration of at least some of these costs might obtain from the implementation of a biophysical economy. Potential reductions in the costs of natural disasters should rightfully enter any benefit:cost analysis undertaken to justify an economic transition to a sustainable economy. As suggested in Chapter 9, data are becoming increasingly available (unfortunately, in part because of increasing natural disasters) to estimate the economic costs of failing to take measurable actions to redress climate change impacts.

Estimating the costs of a transition to a biophysical economy to historical financial institutions and instruments (e.g., saving, lending, liquid investments, commodity markets, etc.) is more difficult. It is recognized that current investments in the stock market and other such liquid instruments are made with the expectation of economic growth and anticipation of sharing in profits and dividends. If a biophysical economy requires or imposes unrealistic limits (real or perceived) on investment earnings, the feasibility or viability of such an economic model might be seriously challenged by profit seekers. As a counterpoint, a biophysical economy might prove less volatile than conventional markets and attract longer-term investors who perceive reduced risk at the expense of unsustainable gains in the near term. The importance of these kinds of instruments for investment in retirement portfolios, IRAs, and 401K plans for millions of individuals in the United States underscores the critical need to reduce risks and secure these investments while transitioning to a biophysical economy. The risk of failing these investors could result in policy constraints that would all but stop such an economic transition.

Any economy, including a biophysical economy, requires finance in the broadest sense to function. Corresponding financial challenges that require solutions in the transition to a sustainable economy include the costs associated with the design of such an economy followed by its initial implementation and subsequent day-to-day operation. An underlying hypothesis is that existing financial institutions and practices can support a BPE or be readily adapted to meet the challenges of a bourgeoning BPE. The testing or evaluation of this hypothesis lies outside the scope of this current presentation, however.

13.7 Addressing Energy Needs

Energy in sufficient quantity and quality from economical and reliable sources is a necessary condition for a sustainable economy – regardless of underlying economic assumptions, biophysical or otherwise.

The concept of energy appears relatively straightforward. However, defining energy poses several conceptual challenges (Giampietro and Sorman 2012). Energy, in physics, is defined as the capacity to do work, for example, exerting a force that displaces an object. Energy exists in many forms, including kinetic, potential, thermal, nuclear, biochemical, and chemical.

A comprehensive approach to evaluating energy transformation, as identified in the Turner et al. model (Figure 13.1), will quantify the contributions of primary energy sources, energy carriers, and end uses of energy (Giampietro and Sorman 2012). These authors recognize the importance of overlapping biophysical constraints in influencing the feasible mix of fossil fuels, nuclear power, and renewable primary energy sources (Figure 13.2). Designing and implementing a BPE will require derivation of a sustainable portfolio of primary energy sources and corresponding energy carriers to sustain end uses of energy, including household, service government, building and manufacturing, and agriculture, forest, and fisheries (Figure 13.2).

13.7.1 Energy Return on Investment

The concept of energy return on investment (EROI) and its application in evaluating alternative energy portfolios will play key roles in the design and operation of a sustainable BPE (Hall and Klitgaard 2018). EROI can be simply described by the equation

EROI = energy returned to society/energy required to get that energy

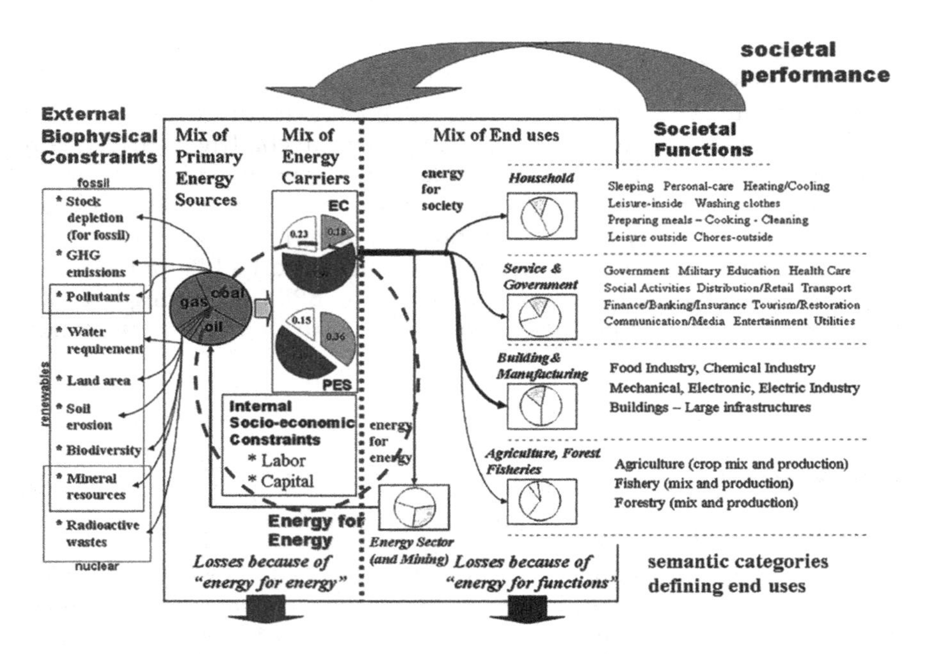

FIGURE 13.2
Energy transformation within societies (Giampietro and Sorman 2012).

The ratio is typically unitless because the units in the numerator and denominator are the same (e.g., joules, Kcals, etc.). A ratio of 10:1 simply means that for an investment of 1 unit of energy, 10 units of energy are produced. Clearly, higher ratios identify efficient energy production. A ratio <1 implies a net loss of energy in the overall return to society. In practice, a Law of Minimum EROI suggests that a ratio of about 3:1 determined for oil and corn-based ethanol at the point of production is necessary for a sustainable society (Hall et al. 2009).

The EROI usually estimates the energy required to get to some point in society, usually the point of production (e.g., wellhead). However, there have been modifications to the above equation to define more precise applications depending on details that bound or locate the calculation. For example, a point of use EROI can be computed as

$EROI_{pou}$ = energy returned to society at point of use/energy required to get and deliver that energy

The design of an energy portfolio for implementing a sustainable economy in the United States can be informed by considering values of EROI for various energy resources, including non-renewable, renewable, and biomass-based technologies (Table 13.2). The selected values underscore the attraction to fossil fuels from purely an energy production perspective. The energy

TABLE 13.2

Values of EROI for Selected Energy Resources and Year in the United States (Taken from Hall and Klitgaard 2018)

Resource	Year	EROI (X:1)
Non-renewable		
Oil and gas (domestic)	1970	30
Oil and gas (domestic)	2007	11
Natural gas	2005	67
Coal (mine mouth)	1950	80
Coal (mine mouth)	2000	80
Coal (mine mouth)	2007	60
Nuclear	n/a	5 to15
Renewable	n/a	
Hydropower	n/a	>100
Wind turbine	n/a	18
Solar flat plate	n/a	1.9
Solar collector	n/a	1.6
Photovoltaic	n/a	6 to 12
Biomass		
Ethanol (sugar cane)	n/a	0.8 to 10
Ethanol (corn-based)	n/a	0.8 to 1.6
Biodiesel	n/a	1.3

returns for investments in oil, natural gas, and coal are typically much greater than for renewable sources (except hydropower). Note that these fossil fuels really represent stocks of concentrated solar energy that have accumulated over millennia. However, as fossil fuel resources diminish over time through extraction, the energy costs of production increase and EROI can correspondingly decrease (e.g., compare oil and gas in 1970 and 2005), ultimately approaching the 3:1 minimum.

As fossil fuel EROI values decrease, these resources become non-viable from purely an energy perspective, regardless of the unit price. However, even at non-viable EROI values, fossil fuels will continue to be extracted and produced to power the US (and global) military for the foreseeable future. To no surprise, the US military is the largest daily consumer of fossil fuels with an annual (2017 data) purchase on the order of 100 million barrels (4.2 billion gallons) of refined petroleum and a daily use in excess of 250,000 barrels or 10.5 million gallons (Morris 2017). For reference, the average daily consumption of petroleum in the United States in 2022 was about 20 million barrels (US Energy Information Administration).

Other than hydropower, renewable resources demonstrate EROI values that do not match up favorably with fossil fuels, except perhaps wind turbine power compared to coal (Table 13.2). Fuels based on biomass appear to perform poorly in terms of EROI with values that are below the 3:1 minimum requirement for a sustainable economy (Hall et al. 2009). The only exception appears to be ethanol derived from sugar cane. Designing and operating a biophysical economy through emphasis on renewable resources will have to address the less energy intense nature of these sources and develop an economy that provides goods and services to satisfy current (and future) demands in the face of more diffuse power sources.

Table 13.2 hints at potentially important feedback among non-renewable and renewable resource utilization. For example, hydropower demonstrates the greatest EROI with values >100. However, the potential for fossil fuel combustion to increase atmospheric CO_2 concentrations with corresponding changes in local and regional hydrology and precipitation could impact hydropower generation. Increased drought compared will, of course, reduce opportunities for hydropower generation; similarly, excessive rainfall at the wrong time could lead to dam releases that bypass generators or add hydropower to the grid when it is not readily accommodated.

Addressing energy needs in building a biophysical economy will need to recognize the scale-dependence of producing or using energy resources in addition to comparing alternative technologies (Figure 13.3). The plot

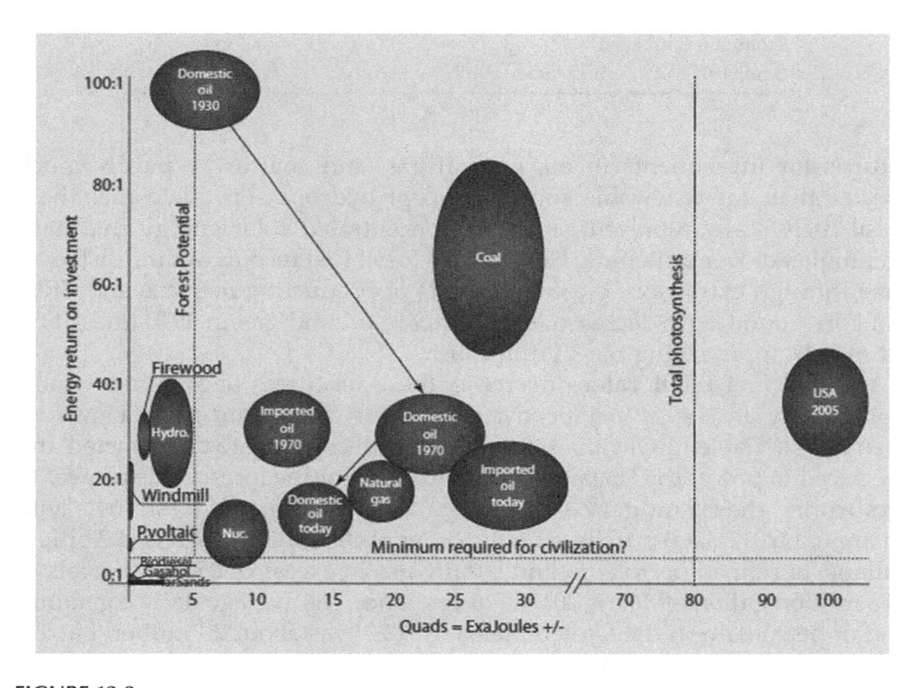

FIGURE 13.3
Balloon graph that illustrates the energy return on investment for various sources and times. The size of each balloon indicates uncertainty associated with the estimate (adapted from Hall and Klitgaard 2018).

underscores the reduction in EROI values for domestic oil from 1930 to 1970 to the present as sources became increasingly energy-intensive to produce (e.g., compare energetic costs of 1930 Oklahoma oil fields to present day offshore deepwater oil extraction). It is possible for the temporal trend to reverse; for example, increased efficiencies in wind and solar technologies might correspondingly increase EROI values through time as technologies presumably advance. The key point is that EROI values are not static, but they reflect changes in resource availability (i.e., peak oil) and changes in the energy required to produce and deliver energy to its end uses. These kinds of dynamics will have to be quantified and incorporated into the development (and periodic updating) of energy technology portfolios designed to support a sustainable biophysical economy in the United States – and eventually worldwide.

The discussion has thus far focused on energy and EROI. However, the development of a sustainable economy will require similar "cost-benefit" assessments of other components of a complex, evolving socioeconomic–environmental complex that defines a biophysical economy. For example, similar concepts and analysis can be developed to evaluate alternative approaches and technologies to manage carbon in relation to climate change. Quantifying the carbon costs per unit carbon sequestered or directly removed might inform management concerning the viability of proposed carbon management schemes and productive use of carbon management technologies – particularly at scale. Recall that it takes about 2.4 billion metrics tons of carbon removed from the atmosphere and stored outside the globally active carbon pools to reduce atmospheric CO_2 by about 1 ppm (Oak Ridge National Laboratory, CDIAC, Carbon Dioxide Information Analysis Center (ornl.gov)).

13.8 A Beta-Version Biophysical Economy

Given the various discussions in preceding chapters, the fundamental questions remain: is it possible to transition from the neoclassical economic model, which emphasizes continuous growth, to a model more closely aligned with finite and renewable resources? If such a transition is possible, how might such an economy be structured? How would corresponding economic functions compare to those that characterize current economies? How might the transition to a biophysical economy manifest?

The proposed biophysical economy is essentially a planned economy. There have been planned economies in the past (e.g., former Soviet Union) and one might argue that the current economy of China is planned to a greater or lesser extent. A key challenge in moving from a laissez-faire capitalist model to more of a planned economy will be maintaining (at least the illusion of) unbridled opportunity and entrepreneurship, while constraining growth to magnitudes compatible with rates of resource renewal or replacement. No small challenge!

There appear to be sufficient private and governmental economic, social, political, and environmental organizations in place to facilitate or at least begin a transition to a biophysical economy. A sustainable economy should emanate through the integration of social, economic, and environmental institutions and processes, as outlined in Figure 13.4. Sustainability is defined as the intersection of these overlapping multi-dimensional organizations, attributes, and functionalities.

Replacing the current mainstream neoclassical economic paradigm with one more closely aligned with the principles of biophysical economics appears as a practical solution in developing a flourishing economy and sustainably inhabiting the planet. The need for a biophysical economy has been well justified from a technical perspective for decades (Hall and Klitgaard 2018; Odum 1971). It is possible, as demonstrated in the proceeding chapters, to prescribe an economy that is ostensibly sustainable. Importantly, the

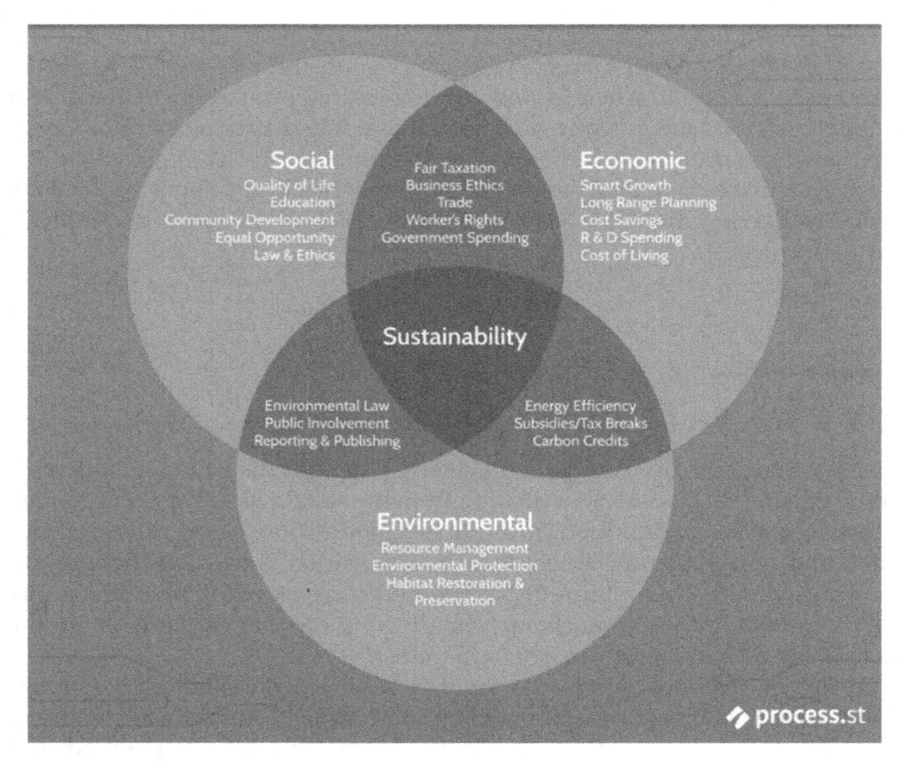

FIGURE 13.4
Key attributes of a sustainable biophysical-based economic model (after process.st).

biogeochemical processes that both sustain life and pose limits to growth are not negotiable. A carrying capacity will be imposed on Earth's human population. Humans to some extent can participate in determining the nature and magnitude of their carrying capacity.

Ideally, a prescription for building a biophysical economy would be based on the conceptual model for a sigmoid approach to some equilibrium (Chapter 11, Figure 11.1 b). Metrics would be defined for selected socioeconomic and environmental dimensions of a sustainable economy, and target values that define sustainable conditions would be derived based on data collected by various government and private agencies. Actions would be undertaken and policies enacted to transform the economy from the current neoclassical model to one consistent with the equilibrium concept. This process would play out in an adaptive management framework, where feedback from monitoring the selected metrics would be used to assess performance in relation to the equilibrium target metrics. The ideal model assumes complete knowledge of the system and its response to manipulations towards sustainability. Target values are static and reflect the aims and goals of a carefully planned sustainable economy in equilibrium. This idealized approach also assumes perfect information from monitoring and absence of significant time lags between a manipulation of the system, its measurable response, and corresponding decision making to continue the adaptive management process.

A more realistic prescription is illustrated by the overshoot and oscillation model (Chapter 11, Figure 11.1 c). This model adopts the same adaptive management approach as outlined above, but it recognizes that there is variability and uncertainty associated with the values of the target metrics. The system is incompletely understood and there is uncertainty concerning the anticipated response of the system to management actions and policies aimed at sustainability. Variability in the monitoring data and potentially ambiguous system responses to management lead to a concept of dynamic equilibrium where management can result in overshoot (or undershoot) – perhaps oscillate along one or many dimensions in relation to moving target values (Figure 11.1 c). Time delays in collecting and processing data, with subsequent delays in application to decision making further lend to the dynamics of adaptively managing the system towards sustainability. Bringing the best science to bear on the economic transformation to a biophysical model will very likely result in a multi-dimensional analog to Figure 11.1 c.

Both the static and the dynamic equilibrium models for sustainability imply the creation of the necessary infrastructure to perform adaptive management and a demonstrated political will among decision makers to execute the necessary socioeconomic and environmental transformations to achieve a sustainable, biophysical economy. This is, of course, possible in theory.

The human experience runs counter to the idealistic models just outlined in relation to economic transformation. The quest to design and implement an economic model that is more aligned with finite planetary resources creates a wicked (in the sense of Lonngren and van Poeck 2021; Termeer et al.

2019; DeFries and Nagendra 2017), maybe even a "super wicked" (Peters 2017) problem – one that has no clear-cut solution and is subject to polarizing and seemingly irreconcilable divergence of opinion and subject to traps of insufficient action or no action at all in the face of apparently hopeless complexity and uncertainty.

Concentrated forms of energy (e.g., woody biomass, fossil fuels, nuclear fission) and a corresponding evolution of technology have allowed humans to construct their own niche and effectively break ecological constraints typically encountered by other species (Waring et al. 2023; Hall and Klitgaard 2018; Allen and Starr 1982). Breaking through ecological constraints has placed humans effectively "outside" their ecosystem, where humans negatively impact their biophysical life-support systems. Remaining constraints on human population dynamics appear at successively larger scales, with current constraints operating at regional to global levels (Nordhaus et al. 2012; Rockstrom et al. 2009). It remains to be demonstrated that the species can effectively manage constraints imposed at global scale. Global-scale problems will require global-scale solutions (Waring et al. 2023; Allen et al. 1999); now is not the time for ultra-nationalism, which is reemerging worldwide. Additionally, it remains to be demonstrated that *H. sapiens* is sufficiently evolved to act upon its own behalf, particularly on a global scale. Thousands of years of continued tribalism, wars, genocide, and other forms of self-inflicted misery argue against the likelihood of successful international collective efforts required to address the challenges of climate change and other global stressors (e.g., reliable sources of potable water and food). Competitive exclusion, as suggested by Mykleby et al. (2016), is not the ecological principle to define a path towards a fair and flourishing nation-scale or global economy. Failure to cooperate and act effectively at scale portends existential risk (Bradshaw et al. 2021; Wallace-Wells 2017; Bostrom 2013).

The problem becomes exceedingly wicked because, as cognitive beings, humans pursue aspirations that know no bounds and have chosen to embrace "infinite human resourcefulness" and "derecognize finitude" as instruments to guide economic growth and development (Costea et al. 2007). Knowledge will continue to increase for as long as there are minds capable of thought, creativity, and innovation (Bostrom 2009). The economic model that harmonizes with human experience is described by the continuous growth model (Figure 11.1 a) – and humans will strive to maintain this model at virtually all costs. Yet the material manifestations of boundless mentality, creation, and innovation remain subject to the physical laws of the Universe, as we understand them, and limitations imposed by finite planetary resources. Historically, when confronted by limitations, capitalists were able to escape to new geographies (i.e., the New World) that afforded virtually untapped resources (Moore 2015). Such an escape in geography seems no longer possible – unless it is to other planets, which resurfaces questions of available energy to make such a journey *en masse*. So, how does one foster imagination and creativity, which are vital to and define human experience, within a highly planned economy that recognizes and imposes material limits to manifesting thought and innovation? The promise of sustainability may prove insufficient.

An alternative prescription for the design and implementation of a biophysical (sustainable) economy follows from the concerns and questions just expressed. The recipe simply calls for continuing with the neoclassical model as the guiding economic paradigm, while addressing the seemingly unavoidable impacts of the continuous growth model using best available technologies. The biophysical limitations will impose themselves as planetary boundaries are exceeded and reconfigure culturally constructed niche space. The habitable and economically viable planet will become smaller and perhaps further challenge future economic opportunity. One advantage of this prescription is that all the necessary institutions and processes to continue the current neoclassical economy are in place and operating. All that is needed is continued advances in technology to redress, as much as possible, planetary impositions on economic growth and development.

The first two prescriptions for sustainable economic development were based on the system dynamics illustrated in Figure 11.1 b and c that describe achieving some kind of equilibrium condition (static or dynamic). The above prescription for a nature-based imposition of sustainability suggest an aperiodic fluctuation in human condition as the resulting system dynamic (Bostrom 2009) – the oscillations indicated within the grey bands in Figure 13.5.

The economic and general quality of life conditions for humans would depend on where the system was in terms of the aperiodic fluctuations defined, in part, by the success of technological advances in ameliorating or mitigating the pressing planetary challenges of the period. The concepts and processes of creative destruction (Diamond 2006) and recurrent Kondratieff cycles (Batty 2015) may become increasingly important in navigating the economy through the peaks and troughs of a planetary-imposed biophysical economy.

One possible positive outcome of this recipe for continued economic growth in the face of planetary challenges is breaking through to a post-human condition as a continuation and consequence of moving through a period of peak technological development – not illustrated in Figure 13.5. A breakthrough of this significance implies advances in technological capabilities sufficient to effectively counter any challenge posed by planetary biophysics. Post-human conditions might be characterized by technical and socioeconomic circumstances that maximize individual well-being and preclude future downturns to historical human conditions – this might be perceived as achieving eudaimonia. Infinite human resourcefulness will have carried the day.

One possible negative outcome of the recipe based on pursuing continuous economic growth is that increasing planetary challenges resulting from human economic activities greatly surpass human technological capabilities

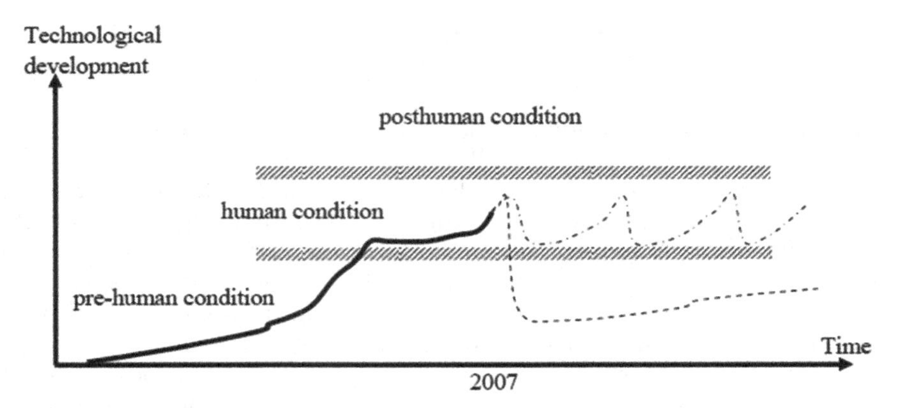

FIGURE 13.5
Possible trajectories of the human condition (from Bostrom 2009).

to remedy them. Limits to growth are imposed to the extent that human extinction becomes a reality. Declining technological prowess, in turn, drives human conditions to pre-human circumstances as a continuation of a period of downturn (Figure 13.5). Sustainability manifests absent human presence.

References

Acemoglu, D., Robinson, J.A. 2013. Economics versus politics: pitfalls of policy advice. *Journal of Economic Perspectives* 27:173–192.

Alcott, B. 2005. Jevons' paradox. *Ecological Economics* 54:9–21.

Allen, T.F.H., Starr, T.B. 1982. *Hierarchy*. The University of Chicago Press, Chicago, IL.

Allen, T.F.H., Tainter, J.A., Hoekstra, T.W. 1999. *Supply-side sustainability*. Columbia University Press.

Alzahrani, F., Collins, A.R., Erfanian, E. 2019. *Drinking water quality impacts on health care expenditures in the United States*. Regional Research Institute Working Papers. 193. https://researchrepository.wvu.edu/rri_pubs/193 West Virginia University.

Armstrong, A. 2020. Ethics and ESG. *AABFJ* 14:1–17.

Backhouse, R.E., Medema, S.G. 2009. On the definition of economics. *Journal of Economic Perspectives* 23:221–233.

Bajaj, A., Ram, S. 2007. A comprehensive framework towards information sharing between government agencies. *International Journal of Electronic Government Research* 3:29–44.

Bakshi, B.R., Fiksel, J. 2003. The quest for sustainability: challenges for process systems engineering. *AIChE Journal* 49:1350–1358.

Banerjee, O., Cicowiez, M., Horridge, M., Vargas, R. 2016. *A conceptual framework for integrated economic-environmental modeling*. Documento de Trabajo, No. 22. Universidad Nacional de la Plata, Laborales y Sociales (CEDLAS), La Plata.

Bar-Lev, D., Katz, S. 1976. A portfolio approach to fossil fuel procurement in the electric utility industry. *The Journal of Finance* 31:933–947.

Bastianoni, S., Pulselli, R.M., Pulselli, F.M. 2009. Models of withdrawing renewable and non-renewable resources based on Odum's energy systems theory and Daly's quasi-sustainability principle. *Ecological Modelling* 220:1926–1930.

Batty, M. 2015. *Creative destruction, long waves, and the age of the smart city*. UCL Working Paper Series 200–Jun 15. University College London. 19 p.

Beaussier, T., Caurla, S., Bellon-Maurel, V. 2019. Coupling economic models and environmental assessment methods to support regional policies: a critical review. *Journal of Cleaner Production* 216:408–421.

Bierman, F., multiple co-authors. 2012. Transforming governance and institutions for global sustainability: key insights from the earth system governance project. *Current Opinion in Environmental Sustainability* 4:51–60.

Binnet, T., Failler, P., Chavance, P.N., et al. 2013. First international payment for marine ecosystem services: the case of Banc d'Arguin National Park, Mauritania. *Global Environmental Change* 23:1434–1443.

Birnbaum, H.G., Carley, C.D., Desai, U., Ou, S., Zuckerman, P.R. 2020. Measuring the impact of air pollution on health care costs. *Health Affairs* 39:2113–2119.

Borner, J., Wunder, S., Reimer, F., et al. 2013. Promoting forest stewardship in the-Bolsa Floresta programme: local livelihood strategies and preliminary impacts. *Global Environmental Change* 30:31–42.

Bostrom, N. 2013. Existential risk prevention as global priority. *Global Policy* 4:15–31.

Bostrom, N. 2009. The future of humanity, pp. 186–216. In Olsen, J.-K.B., Selinger, E., Riis, S. (eds.), *New waves in philosophy of technology*. Palgrave MacMillan, New York.

Bostrom, N. 2002. Existential risks: analyzing human extinction scenarios and related hazards. *Journal of Evolution and Technology* 9:1–36.

Bouslah, K., Kryzanowski, L., M'Zali, B. 2013. The impact of the dimensions of social performance on firm risk. *Journal of Banking & Finance* 37:1258–1273.

Bradshaw, C.J.A., multiple co-authors. 2021. Underestimating the challenges of avoiding a ghastly future. *Frontiers in Conservation Science* 1:615419. http://dx.doi.org/10.3389/fcosc.2020.615419

Brodie, J. 2014. Dredging the great barrier reef: use and misuse of science. *Estuaries, Coastal and Shelf Sciences* 142:1–3.

Broecker, W.S. 2002. Constraints on the glacial operation of the atlantic ocean's conveyor circulation. *Israel Journal of Chemistry* 42:1–14.

Brussard, P.F., Reed, J.M., Tracy, C.R. 1998. Ecosystem management: what is it really? *Landscape and Urban Planning* 40:9–20.

Canfield, D.E., Glazer, A.N., Falkowski, P.G. 2010. The evolution and future of Earth's nitrogen cycle. *Science* 330:192–196.

Chang, C.-P., Zhang, W.-L. 2020. Do natural disasters increase financial risks? An empirical analysis. *Bulletin of Monetary Economics and Banking*. Special Issue 2020:61–81.

Chapin, III, F.S., Kofinas, G.P., Folke, C. 2009. *Principles of ecosystem stewardship*. Springer, New York.

Christensen, N.L., Bartuska, A.M., Brown, J.H., Carpenter, S., D'Antonio, C., Francis, R., Franklin, J.F., MacMahon, J.A., Noss, R.F., Parsons, D.J., Peterson, C.H., Turner, M.G., Woodmansee, R.G. 1996. The report of the ecological society of America. Committee on the scientific basis for ecosystem management. *Ecological Applications* 6: 665–691.

Cianci, R., Gambrel, P. 2003. Maslow's hierarchy of needs: does it apply in a collectivist culture. *Journal of Applied Management and Entrepreneurship* 8:143–161.

Cleveland, C.J., Costanza, R., Hall, C.A., Kaufmann, R. 1984. Energy and the US economy: a biophysical perspective. *Science* 225:890–897.

Cohen, J.E. 2017. How many people can the Earth support? *Journal of Population and Sustainability* 2:37–42.

Cohen, J.E. 1997. Population, economics, environment and culture: an introduction to human carrying capacity. *Journal of Applied Ecology* 34:1325–1333.

Cohen, J.E. 1995. Population growth and Earth's carrying capacity. *Science* 269:341–346.

Cornell, B., Damodaran, A. 2020. Valuing ESG: doing good or sounding good? *Journal of Impact and ESG Investing* 1:76–93. https://doi.org/10.3905/jesg.2020.1.1.076.

Cosenz, F., Bivona, E. 2021. Fostering growth patterns of SMEs through business model innovation. A tailored dynamic business modeling approach. *Journal of Business Research* 130:658–669.

Cosenz, F., Rodrigues, V.P., Rosati, F. 2019. Dynamic business modelling for sustainability: exploring a system design perspective. 26th EurOMA Conference Operations Adding Value to Society, pp. 999–1004.

Cosenz, F., Rodrigues, V.P., Rosati, F. 2020. Dynamic business modeling for sustainability: exploring a systems dynamics perspective to develop sustainable business models. *Business Strategy and the Environment* 29:651–664.

Costanza, R. 2010. Toward a new sustainable economy. *Real-World Economics Review* 49:20–21.

Costanza, R. 1980. Embodied energy and economic valuation. *Science* 210:1219–1224.

Costanza, R., dArge, R., de Groot, R., Farber, S., Grasso, M., Hannon, B., Limburg, K., Naeem, S., O'Neill, R.V., Paruelo, J., Raskin, R.G., Sutton, P., van den Belt, M. 1997. The value of the world's ecosystem services and natural capital. *Nature* 387:253–260.

Costanza, R., De Groot, R., Braat, L., Kubiszewski, I., Fioramonti, L., Sutton, P., Farber, S., Grasso, M. 2017. Twenty years of ecosystem services: how far have we come and how far do we still need to go? *Ecosystem Services* 28:1–16.

Costanza, R., Graumlich, L., Steffen, W., Crumley, C., Dearing, J., Hibbard, K., Leemans, R., Redman, C., Schimel, D. 2007. Sustainability or collapse: what can we learn from the history of humans and the rest of nature? *Royal Swedish Academy of Science* 36:522–527.

Costanza, R., Leemans, R., Boumans, R., Gaddis, E. 2007. Integrated global models, pp. 417–446. In Costanza, R., Graumlich, L.J., Steffen, W. (eds.), *Sustainability or collapse: an integrated history and future of people on earth*. Dahlem Workshop Report 96. MIT Press, Cambridge, MA.

Costea, B., Crump, N., Amiridis, K. 2007. Managerialism and *'infinite human resourcefulness'*: a commentary upon the *'therapeutic habitus'*, *'derecognition of finitude'* and the modern sense of self. *Journal for Cultural Research* 11:1–21.

Crimmins, A.R., Avery, C.W., Easterling, D.R., Kunkel, K.E., Stewart, B.C., Maycock, T.K. (eds.). 2023. *Fifth national climate assessment*. US Global Change Research Program. Washington, DC. https://doi.org/10.7930/NCA5.2023.

CTI. 2012. *Designing coastal and marine-based payment for ecosystem services*. Coral Triangle Initiative on Coral Reefs, Fisheries and Food Security, Jakarta, Indonesia. http://www.coraltriangleinitiative.org/library/financial-priority-designing-coastal-and-marine-based-payment-ecosystemservices-pes.

Daepp, M.I.G., Hamilton, M.J., West, G.B., Bettencourt, L.M.A. 2015. The mortality of companies. *Journal of the Royal Society Interface* 12:20150120. http://dx.doi.org/10.1098/rsif.2015.0120.

Daily, G.C. 2000. Management objectives for the protection of ecosystem services. *Environmental Science and Policy* 3:333–339.

Daly, H.E. 1977. *Steady-state economy*. Freeman Press, San Francisco.

DeFries, R., Nagendra, H. 2017. Ecosystem management as a wicked problem. *Science* 356:265–270.

deLlano-Paz, F., Calvo-Silvosa, A., Antelo, S.I., Soares, I. 2017. Energy planning and modern portfolio theory: a review. *Renewable and Sustainable Energy Reviews* 77:636–651.

Delucchi, M.A. 2004. Summary of the non-monetary externalities of motorvehicle use. Report #9 in the series (Revised version). The Annualized Societal Cost of Motor-Vehicle Use in the United States Based in 1990e1991 Data. Research Report UCD-ITS-RR-96-03. Institute of Transportation Studies, University of California, Davis, CA.

Deutsche Bank. 2024. *Sustainable finance framework*. Deutsche Bank Aktiengesellschaft, Frankfurt am Main, Germany.

Diamond, A.M., Jr. 2006. Schumpeter's creative destruction: a review of the evidence. *Journal of Private Enterprise* 22:120–146.

Diamond, J. 2005. *Collapse: how societies choose to fail or succeed.* Viking Press, New York.

Dima, M., Lohmann, G. 2010. Evidence of two distinct modes of large-scale ocean circulation changes over the last century. *American Meteorological Society* 23:5–1.

Dobbs, T.L., Pretty, J. 2008. Case study of agri-environmental payments: The United Kingdom. *Ecological Economics* 65:765–775.

Doury, A., Somot, S., Gadat, S., Ribes, A., Corre, L. 2022. Regional climate model emulator based on deep learning: concept and first evaluation of a novel hybrid downscaling approach. *Climate Dynamics.* https://doi.org/10.1007/s00382-022-06343-9

Drupp, M.A., Hansel, M.C., Fenichel, E.P., Freeman, M., Gollier, C., Groom, B., Heal, G.M., Howard, P.H., Millner, A., Moore, F.C., Nesje, F., Quaas, M.F., Smulders, S., Sterner, T., Traeger, C., Venmans, F. 2024. Accounting for the increasing benefits from scarce ecosystems. *Science* 383:1062–1064.

EAFPES. 2013. *A small scale carbon-offset project in mangroves.* East Africa Forum for Payment of Ecosystem Services. http://www.eafpes.org/index.php/researc handpublications/extensions/mikoko-pamoja-outline.

Eccles, R.G., Ioannou, I., Serafeim, G. 2014. The impact of corporate sustainability on organizational processes and performance. *Management Science* 60:2835–2857.

Eckelman, M.J., Huang, K., Lagasse, R., Senay, E., Dubrow, R., Sherman, J.D. 2020. Health care pollution and public damage in the United States: an update. *Health Affairs* 39:2071–2079.

Ecoinvent Centre. 2004. Ecoinvent data v1.2. Final reports ecoinvent 2000, vol. 1–15, Swiss Centre for Life Cycle Inventories, Dubendorf.

Edwards PN. 1996. Global comprehensive models in politics and policy making. *Climatic Change* 32:149–161.

Federowicz, J., Gogan, J.L., Culnan, M.J. 2010. Barriers to interorganizational sharing ine-Government: a stakeholder analysis. *The Information Society* 26:315–329.

Forest Trends. *Oddar meanchey REDD project (Cambodia).* Forest Carbon Portal, Forest Trends, Washington, DC. http://www.forestcarbonportal.com/project/oddar -meanchey forest-carbon-project

Franchini, M., Mannucci, P.M., Harari, S., Pontoni, F., Croci, E. 2015. The health and economic burden of air pollution. *The American Journal of Medicine* 128:931–932.

Friess, D.A., Phelps, J., Garmendia, E., Gomez-Baggerthun, E. 2015. Payments for ecosystem services (PES) in the face of external biophysical stressors. *Global Environmental Change* 30:31–42.

Frischknecht, R. Jungbluth, R., Althaus, N., Doka, H-J., Dones, G., Heck, R., Hellweg, S., Hischier, R., Nemecek, T., Rebitzer, G., Spielmann, M. 2005. The ecoinvent database: overview and methodological framework. *International Journal of Life Cycle Assessment* 10:3–9.

Georgescu-Roegen, N. 1971. *The entropy law and the economic process.* Harvard University Press, Cambridge.

Giampietro, M., Mayumi, K. 2018. Unraveling the complexity of Jevons paradox: the link between innovation, efficiency, and sustainability. *Frontiers in Energy Research* 6:26. https://doi.org/10.3389/fenrg.2018.00026.

Giampietro, M., Sorman, A.H. 2012. Are energy statistics useful for making energy scenarios? *Energy* 37:5–17.

Giorgi, F. 2019. Thirty years of regional climate modeling: where are we and where are we going next? *Journal of Geophysical Research: Atmospheres* 124: 5696–5723. https://doi.org/10.1029/2018JD030094

Goldin, I., Vogel, T. 2010. Global governance and systemic risk in the 21st century: lessons from the financial crisis. *Global Policy* 1:1 http://dx.doi.org/10.1111/j.1758-5899.2009.00011.x

Gross-Camp, N., Martin, A., McGuire, S., et al. 2012. Payments for ecosystem services in an African protected area: exploring issues of legitimacy, fairness, equity and effectiveness. *Oryx* 46: 24–33.

Grosskurth, J. 2007. Ambition and reality in modeling: a case study on public planning for regional sustainability.

Grumbine, R.E.1994. What is ecosystem management? *Conservation Biology* 8:27–38.

Hall, C.A.S., Balogh, S., Murphy, D.J.R. 2009. What is the minimum EROI that a sustainable society must have? *Energies* 2:25–47.

Hall, C.A.S., Cleveland, C., Berger, M. 1981. Energy return on investment for United States petroleum, coal and uranium. In: Mitch, W. (ed.), *Energy and Ecological Modeling*, Symposium Proceedings, 715–724.

Hall, C.A.S., Cleveland, C.J., Kaufmann, R., 1986. *Energy and resource quality: the ecology of the economic process.* Wiley-Interscience, New York.

Hall, C.A.S., Klitgaard, K. 2019. The need for, and growing importance of, biophysical economics. *Current Analysis of Economics & Finance* 1:75–87.

Hall, C.A.S., Klitgaard, K. 2018. *Energy and the wealth of nations.* Springer, Cham, Switzerland, 511 p.

Hall, C.A.S., Lavine, M., Sloane, J. 1979. Efficiency of energy delivery systems: part I. An economic and energy analysis. *Environmental Management* 3(6):493–504.

Harte, J. 2001. Land use, biodiversity, and ecosystem integrity: the challenge of preserving Earth's life support system. *Ecology Law Quarterly* 27:929–964.

Henisz, W., Koller, T., Nuttall, R. 2019. *Getting your environmental, social, and governance (ESG) proposition right links to higher value creation. Here's why.* McKinsey & Company. Chicago, Illinois, 12 pp.

Herrington, G. 2022. Update to limits to growth. *Journal of Industrial Ecology* 2020:1–13.

Higgins, R.C. 1977. How much growth can a firm afford? *Financial Management* 6:7–16.

Holman, J.O. 2020. Consumer ecoregions: geographic segmentation of the western conterminous United States. *Journal of Applied Business and Economics* 22 (11).

Holman, J.O., Hacherl, A. 2021. Consumer ecoregions: geographic segmentation of the eastern conterminous United States of America. *Journal of Applied Business and Economics* 23:161–170.

Hotelling, H. 1991. The economics of exhaustible resources. *Bulletin of Mathematical Biology* 53:281–312.

Hughes, B.B. 2019. *International futures: building and using global models.* Academic Press, London, UK.

Hughes, B.B. 2009. *Forecasting long-term global change: introduction to International Futures (IFs).* Frederick S. Pardee Center for International Futures, University of Denver. www.ifs.du.edu.

Huijbregts, M.A.J., Hellwig, S., Frischknecht, R., Hungerbuhler, K., Hendriks, A.J. 2008. Ecological footprint accounting in the life cycle assessment of products. *Ecological Economics* 64:798–807.

IEA (International Energy Agency). 2019. World energy outlook 2019. www.iea.org/reports/world-energy-outlook-2019

Isaac, A.G. 1998. The structure of neoclassical consumer theory. (Preliminary draft).

Ivanova, M. 2011. *Global governance in the 21st century: rethinking the environmental pillar.* Conflict Resolution, Human Security, and Global Governance Faculty Publication Series 1. https://scholarworks.umb.edu/crhsgg_faculty_pubs/1

Iversen, T., Soskice, D. 2020. *Democracy and prosperity: reinventing capitalism through a turbulent century.* Princeton University Press. Princeton, New Jersey.

Jacques, P., Delannoy, L., Andrieu, B., Yilmaz, D., Jeanmart, H., Godin, A. 2023. Assessing the economic consequences of an energy transition through a biophysical stock-flow consistent model. *Ecological Economics* 209:107832. https://doi.org/10.1016/j.ecolecon.2023.107832

Ji, X., Luo, Z. 2020. Opening the black box of economic processes: ecological economics from its biophysical foundation to a sustainable economic institution. *The Anthropocene Review* 7:231–247.

Katsamakas, E., Miliaresis, K., Pavlov, O.V. 2022. Digital platforms for the common good: social innovation for active citizenship and ESG. *Sustainability* 14:639. https://doi.org/10.3390/su14020639.

Kering. 2013. *Kering environmental profit & loss. Methodology and 2013 group results.* Kering. Paris, France.

Kim, S., Li, Z. 2021. Understanding the impact of ESG practices in corporate finance. *Sustainability* 13:3746. https://doi.org/10.3390/su13073746.

King-Hill, S. 2015. Critical analysis of Maslow's hierarchy of need. *The STeP Journal* 2:54–57.

Kissinger, M., Sussman, C., Moore, J., Rees, W.E. 2013. Accounting for the ecological footprint of materials in consumer goods at the urban scale. *Sustainability* 5:1960–1973.

Kitzes, J., Wackernagel, M. 2009. Answers to common questions in ecological footprint accounting. *Ecological Indicators* 9:812–817.

Klitgaard, K. 2020. Sustainability as an economic issue: a biophysical economic perspective. *Sustainability* 12:364. https://doi.org/10.390/su120110364.

Koehler, B., Abulafia, D., Bateman, V., Bowen, H., Crafts, N. 2015. *An introduction to the history of capitalism 600 – 1900 AD.* Legatum Institute. www.li.com

Konduri, V.S., Kumar, J., Hargrove, W.H., Hoffman, F.M., Ganguly, A.R. 2020. Mapping crops within the growing season across the United States. *Remote Sensing of Environment* 251:112048 https://doi.org/10.1016/j.rse.2020.112048

Kosoy, N., Corbera, E., Brown, K. 2008. Participation in payments for ecosystem services: case studies from the Lacandon rainforest, Mexico. *Geoforum* 39:2073–2083.

Kotsantonis, S., Pinney, C., Serafeim, G. 2015. ESG integration in investment management: myths and realities. *Journal of Applied Corporate Finance* 28:10–16.

Krautkraemer, J.A. 2005. *Economics of natural resource scarcity: the state of the debate.* Resources for the Future, Discussion Paper 05–14, 1616 P Street, NW, Washington, DC.

Kumar, N.C.A., Smith, C., Badis, L., Wang, N., Ambrosy, P., Tavares, R. 2016. ESG factors and risk-adjusted performance: a new quantitative model. *Journal of Sustainable Finance & Investment* 6:292–300.

Kuznets, S. 1973. Modern economic growth: findings and reflections. *The American Economic Review* 63:247–258.

Lenton, T.M., Held, H., Kriegler, E., Hall, J.W., Lucht, W., Rahmstorf, S., Schellnhuber, H.J. 2008. Tipping elements in Earth's climate system. *Proceedings of the National Academy of Sciences* 105:1786–1793.

Little, W. 2023. *Introduction to sociology.* 3rd Canadian ed., Creative Commons, Nova Scotia, Canada.

Liu, P., Chetal, A. 2005. Trust-based secure information sharing between federal government agencies. *Journal of the American Society for Information Science and Technology* 56:283–298.

Lönngren, J., van Poeck, K. 2021. Wicked problems: a mapping review of the literature. *International Journal of Sustainable Development & World Ecology* 28:481–502.

Lotka, A.J. 1925. *Elements of physical biology*. Williams and Wilkins, Baltimore.

Lotka, A.J. 1922. Contribution to the energetics of evolution. *Proceedings of the National Academy of Sciences* 8:147.

Lozano, R. 2015. A holistic perspective on corporate sustainability drivers. *Corporate Social Responsibility and Environmental Management* 22:32–44.

Maher, P., Gerber, E.P., Medeiros, B., Merlis, T.M., Sherwood, S., Sheshadri, A., et al. 2019. Model hierarchies for understanding atmospheric circulation. *Reviews of Geophysics* 57:250–280.

Malthus, T.R. 1986. *An essay on the principle of population*. 1st ed., of 1798. Pickering, London.

Malthus, T.R. 1798. *An essay on the principle of population*. Johnsson, London.

Margolis, J.D., Elfenbein, H.A., Walsh, J.P. 2009. *Does it pay to be good... and does it matter? A meta-analysis of the relationship between corporate social and financial performance*. Working Paper, SSRN: http://dx.doi.org/10.2139/ssrn.1866371, United States.

Martinet, V., Doyen, L. 2007. Sustainability of an economy with an exhaustible resource: a viable control approach. *Resource and Energy Economics* 29:17–9.

Maslow, A. 1943. A theory of human motivation. *Psychological Review* 50:370–396.

Meadows, D.H., Meadows, D.L., Randers, J. 1992. *Beyond the limits: global collapse or a sustainable future*. Earthscan Publications, Ltd., London.

Meadows, D.H., Meadows, D.L., Randers, J., Behrens III, W.W. 1972. *The limits to growth: a report for the Club of Rome's project on the predicament of mankind*. Universe Books, New York.

Meadows, D.H., Randers, J., Meadows, D.L. 2004. *Limits to growth: the 30 year update*. Chelsea Green Publishing Company, White River Junction, Vermont.

Mearns, L.O., Bogardi, I., Giorgi, F., Matyasovszky, I., Palecki, M. 1999. Comparison of climate change scenarios generated from regional climate model experiments and statistical downscaling. *Journal of Geophysical Research* 104:6603–6621.

Melgar-Melgar, R.E., Hall, C.A.S. 2020. Why ecological economics needs to return to its roots: the biophysical foundation of socio-economic systems. *Ecological Economics* 169:106567 https://doi.org/10.1016/j.ecolecon.2019.106567

Millennium Ecosystem Assessment (MEA). 2005. *Ecosystems and human well-being*. Island Press, Washington, DC.

Mokyr, J. 2007. The market for new ideas and the origins of the economic growth in eighteenth century Europe. *Tijdschrift voor Sociale en Economische Gerschiedenis* 4:3–38.

Mokyr,J. 2005. The intellectual origins of modern economic growth. *The Journal of Economic History* 65:285–351.

Monciardini, D. 2012. Good business? The struggles for regulating ESG disclosures. *Onati Socio-Legal Series* 2:1–23. http://ssrn.com/abstract=2041918

Moore, J.W. 2015. *Capitalism in the web of life – ecology and the accumulation of capital*. Verso, London.

Morris, B.S. 2017. The one thing causing climate change no one wants to talk about. https://medium.com/@brettsmorris/the-one-thing-causing-climate-chang...

Motesharrie, S., multiple co-authors. 2016. Modeling sustainability: population, inequality, consumption, and bidirectional coupling of earth and human systems. *Natural Science Review* 3:470–494.

Motesharrei, S., Rivas, J., Kalnay, E. 2014. Human and nature dynamics (HANDY): modeling inequality and use of resources in the collapse or sustainability of societies. *Journal of Ecological Economics* 101:90–102.

Muller, N.Z., Mendelsohn, R., Nordhaus, W. 2011. Environmental accounting for pollution in the United States economy. *American Economic Review* 101:1649–1675.

Mykleby, M., Doherty, P., Makower, J. 2016. *The new grand strategy: restoring America's prosperity, security, and sustainability in the 21st century.* St. Martin's Press. New York.

Nilsson, C., Grelsson, G. 1995. The fragility of ecosystems: a review. *Journal of Applied Ecology* 32:677–692.

Nordhaus, T., Shellenberger, M., Blomqvist, L. 2012. *The planetary boundaries hypotheses. A review of the evidence.* The Breakthrough Institute, Oakland, CA.

Obst, C.G. 2015. Reflections on natural capital accounting at a national level: advances in the system of environmental-economic accounting. *Sustainability Accounting, Management and Policy Journal* 6:315–339.

Odum, H.T. 1971. *Environment, power, and society.* Wiley-Interscience, New York.

Oikonomou, I., Brooks, C., Pavelin, S. 2012. The impact of corporate performance on financial risk and utility: a longitudinal analysis. *Financial Management* 41:483–515.

Oki, T. 2005. 2: The hydrological cycles and global circulation. In Anderson, M.G. (ed.), *Encyclopedia of hydrological sciences.* Wiley and Sons, Ltd. Hoboken, New Jersey.

Oprean-Stan, C., Oncioiu, I., Iuga, I.C., Stan, S. 2020. Impact of sustainability reporting and inadequate management of ESG factors on corporate performance and sustainable growth. *Sustainability* 12:8536. https://doi.org/10.3390/su12208536.

Orlitzky, M., Schmidt, F.L., Rynes, S.L. 2003. Corporate social and financial performance. A meta-analysis. *Organizational Studies* 24:403–441.

Otajacques, B., Hitzelberger, P., Feltz, F. 2007. Interoperability of E-Government information systems: issues of identification and data sharing. *Journal of Management Information Systems* 23:29–51.

Page, S.E. 2005. Are we collapsing? A review of Jared Diamond's Collapse: how societies choose to fail or succeed. *Journal of Environmental Literature* 43:1049–1062.

Park, S.R., Jang, J. 2021. The impact of ESG management on investment decisions: institutional investors' perception of country-specific ESG criteria. *International Journal of Financial Studies* 9:48. https://doi.org/10.3390/ijfs9030048.

Pavlikaksis, G.E., Tsihrintzis, V.A. 2000. Ecosystem management: a review of a new concept and methodology. *Water Resources Management* 14:257–283.

Perrot-Maıˆtre, D. 2006. *The vittel payments for ecosystem services: a "perfect" PES case?* IIED, London.

Peters, B.G. 2017. What is so wicked about wicked problems? A conceptual analysis and a research program. *Policy and Society* 36:385–396.

Pilon, R. 2017. Property rights and the constitution, pp. 173–191. In Pilon, R. (ed.), *CATO handbook for policy makers.* CATO Institute. Washington DC.

Polimeni, J.M., Mayumi, K., Giampietro, M., Alcott, B. 2008. *The Jevons paradox and the myth of resource efficiency improvements.* Earthscan Publishers, London. 184 p.

Porter, M.E., Kramer, M.R. 2011. *Creating shared value.* Harvard Business.

Poston, B. 2009. Maslow's hierarchy of needs. *The Surgical Technologist,* August 2009, pp. 347–353.

Platt, H.D., Platt, M.B., Chen, G. 1995. Sustainable growth rate of forms in financial distress. *Journal of Economics and Finance* 19:147–151.

PUMA. 2017. *PUMA annual report 2017.* Sustainability. Herzogenaurach, Germany.

Quesnay, F. 1765. Observation sur la droit naturel des hommes reunis en societe. *Journal of Agriculture, Commerce, and Finance,* September 17–38. https://www.marxists.org/reference/subject/economics/quesnay/1765/rights.htm.

Ratcliffe, C., Congdon, W., Teles, D., Stanczyk, A., Martin, C. 2020. From bad to worse: natural disasters and financial health. *Journal of Housing Research* 29(Suppl. 1):S25–S53.

Rathje, S. 2009. The definition of culture: an application-oriented overhaul. *Online-Zeitschrift für interkulturelle Studien* 8(8):35–58. https://nbn-resolving.org/urn:nbn:de:0168-ssoar-455417.

Rees, W.E. 1992. Ecological footprints and appropriated carrying capacity: what urban economics leave out. *Environment and Urbanization* 4:121–130.

Reuters. 2012. Explainer: how four big companies control the U.S. beef industry. June 17, 2021.

Richardson, K.E. and multiple co-authors. 2023. Earth beyond six of nine planetary boundaries. *Science Advances* 9:1–16.

Rockstrom, J., multiple co-authors. 2009. Planetary boundaries: exploring the safe operarting space for humanity. *Ecology and Society* 142:32. http://www.ecologyandsociety.org/vol114/iss2/art32/

Rodrick, D. 2009. *Reinventing capitalism.* International Economy Winter 2009. pp. 78–79.

Romer, P.M. 1996. *Why, indeed, in America? Theory, history, and the origins of modern economic growth.* NBER Working Paper Series, Working Paper 5443, National Bureau of Economic Research, Cambridge, MA. 15 p.

Rosa, H., Kandel, S., Dimas, L. 2004. Compensation for environmental services and rural communities: lessons from the Americas. *International Forestry Review* 6:187–194.

Rosenau, J.N. 1995. Governance in the twenty-first century. *Global Governance* 1:13–43.

Sachs, J.D. 1999. Twentieth-century political economy: a brief history of global capitalism. *Oxford Review of Economic Policy* 15:90–101.

Sassen, R., Hinze, A.-K., Hardeck, I. 2016. Impact of ESG factors on firm risk in Europe. *Journal of Business Economics* 86:867–904.

Schneider, T. 2006. The general circulation of the atmosphere. *Annual Review of Earth and Planetary Science* 34:655–688.

Schellnhuber, H.J. 2002. Coping with Earth system complexity and irregularity, pp. 151–159. In Steffen, W., Jaeger, J., Carson, D.J., Bradshaw, C. (eds.), *Challenges of a changing Earth.* Springer Verlag, Berlin, German.

Schwab, K. 2021. *Stakeholder capitalism: a global economy that works for progress, people, and planet.* Wiley Publishers. Hoboken, New Jersey.

Sherwood, J., Carbajales-Dale, M., Haney, B.R. 2020. Putting the biophysical (back) in economics: a taxonomic review of modeling the earth-bound economy. *Biophysical Economics and Sustainability* 5:4 https://doi.org/10.1007/s41247-020-00069-0

SNA. 2009. *System of National Accounts.* United Nations, New York. 722 p.

Soddy, F. 1922. *Cartesian economics.* Hendersons, London.

Steffen, W. and multiple co-authors. 2015. Planetary boundaries: guiding human development on a changing planet. *Science* 347:1259855-1–1259855-10.

Stein, L.Y., Klotz, M.G. 2016. The nitrogen cycle. *Current Biology* 26:R83–R101.
Sterman, J.D. 2000. *Business dynamics – systems thinking and modeling for a complex world*. Irwin McGraw-Hill, Boston. 982 p.
Stiglitz, J. 1974. Growth with exhaustible natural resources: efficient and optimal growth paths. *The Review of Economic Studies* 41:123–137.
Tay., L., Diener, E. 2011. Needs and subjective well-being around the world. *Journal of Personality and Social Psychology* 31:114–135.
TEEB (The Economics of Ecosystems and Biodiversity). 2010. *Mainstreaming the economics of nature: a synthesis of the approach, conclusions, and recommendations of TEEB*. Earthscan, London.
Tegtmeier, E.M., Duffy, M.D. 2004. External costs of agricultural production in the United States. *International Journal of Agricultural Sustainability* 2:1–20.
Termeer, C.J.A.M., Dewulf, A., Biesbroek, R. 2019. A critical assessment of the wicked problem concept: relevance and usefulness for policy and science. *Policy and Society* 38:167–179.
Tijani, A., Ahmadi, A. 2022. An empirical study of the relationship between the busy outside directors and indicators of ESG performance. *Decision Science Letters* 11:323–332.
Treanor, W.M. 1995. The original understanding of the takings clause and the political process. *Columbia Law Review* 95:782–887.
Turner, G. 2014. *Is global collapse imminent?* MSSI Research Paper No. 4, Melbourne Sustainability Society Institute, The University of Melbourne.
Turner, G. 2008. *A comparison of the limits to growth with thirty years of reality*. CSIRO Working Paper Series 2008-09. Canberra, Australia.
Turner, G.M., Hoffman, R., McInnis, B.C., Poldy, F., Foran, B. 2011. A tool for strategic biophysical assessment of a national economy – the Australian stocks and flows framework. *Environmental Modelling & Software* 26:1134–1149.
Ulrich, C., Vermard, Y., Dolder, P.J., Brunel, T., Jardim, E., Holmes, S.J., Kempf, A., Mortensen, L.O., Poos, J-J., Rindorf, A. 2017. Achieving maximum sustainable yield in mixed fisheries: a managment approach for the North Sea demersal fisheries. *ICES Journal of Marine Science* 74:566–575.
UNDP (United Nations Development Programme). 2015. *Human development index*. World Bank Databank. United Nations, New York, https://hdr.undp.org/data-center/human-development-index#/
United Nations, European Commission and Organisation of Economic Co-operations and Development and Work Bank. 2013. *System of environmental-economic accounting 2012 –experimental ecosystem accounting*. White cover publication, New York, NY.
USEPA (United States Environmental Protection Agency). 1974. *Carrying capacity in regional environmental management*. EPA 600/5-74-021. Washington, DC.
Verstegen, S.W., Hanekamp, J.C. 2005. The sustainability debate: idealism versus conformism – the controversy over economic growth. *Globalizations* 2:349–362.
Vodafone. 2014/2015. *Environmental profit and loss. Methodology and results*. Vodafone, Netherlands.
Wackernagel, M., Rees, B. 1996. *Our ecological footprint: reducing human impact on earth*. New Society Publishers, Gabrioloa Isalnd BC, Canada. 160 p.
Wackernagel, M., Onisto, L., Bello, P., Linares, A.C., Falfan, I.S.L., Garcia, J.M., Guerrero, A.I.S., Guerrero, M.G.S. 1999. National natural capital accounting with the ecological footprint concept. *Ecological Economics* 29:375–390.

Wallace-Wells, D. 2017. The uninhabitable earth. *New York Magazine*, July 10, 2017.
Walton, D., Berg, N., Pierce, D., Maurer, E., Hall, A., Lin, Y.-H., Rahimi, S., Cayan, D. 2020. Understanding differences in California climate projections produced by dynamical and statistical downscaling. *Journal of Geophysical Research: Atmospheres* 12:e2020JD032812. https://doi.org/10.1029/2020JD032812
Wang, W., Javaid, M.U., Bano, S., Younas, H., Jan, A., Salameh, A.A. 2022. Interplay among institutional actors for sustainable economic development – role of green policies, ecoprenuership, and green technological innovation. *Frontiers in Environmental Science* 10:956824. https://doi.org/10.3389/fenvs.2022.956824
Waring T.M., Wood Z.T., Szathmáry E. 2023 Characteristic processes of human evolution caused the Anthropocene and may obstruct its global solutions. *Philosophical Transactions of the Royal Society B* 379: 20220259. https://doi.org/10.1098/rstb.2022.0259
Watt, J. 2024. Just 57 companies linked to 80% of greenhouse gas emissions since 2016. *The Guardian*, April 3, 2024.
West, G. 2017. *Scale - The universal laws of life and death in organisms, cities, and companies*. Weidenfeld & Nicolson, London, UK. 479 p.
Wigley, T.M.L. 2004. *Input needs for downscaling of climate data*. Discussion Paper 500-04-027. California Energy Commission. Sacramento, California.
Wright, R.G., Murray, M.P., Merrill, T. 1998. Ecoregions as a level of ecological analysis. *Biological Conservation* 86:207–213.
Yan, J., Feng, L., Steblyanskaya, Kleiner, G., Rybachuk, M. 2019. Biophysical economics as a new paradigm. *International Journal of Public Administration*. https://doi.org/10.1080/01900692.2019.1645691
York, R., McGee, J.A. 2015. Understanding the Jevons paradox. *Environmental Sociology*. http://dx.doi.org/10.1080/23251042.2015.1106060
Zabel, R.W., Harvey, C.J., Katz, S.L., Good, T.P., Levin, P.S. 2003. Ecologically sustainable yield. *American Scientist* 91:150–157.

Suggested Reading

Ahmad, T., Zhang, D. 2020. A critical review of comparative historical energy consumption and future demand: the story told so far. *Energy Reports* 6:1973–1991.
Batty, M. 2015. *Creative destruction, long waves, and the age of the smart city*. UCL Working Paper Series 200--Jun 15, University College London. 19 p.
Beinhocker, E. 2012. New economics, policy, and politics, pp. 13–146. In Dolphin, T., Nash, D. (eds.), *Complex new world. Translating new economic thinking into public policy. New era economics*. Institute for Public Policy Research.
Cheffins, B.R. 2021. Stop blaming Milton Friedman! Washington University Law Review 98:1607–1644.
Costanza, R. 1981. Embodies energy, energy analysis, and economics, pp. 119–146. In Daly, H.E., Umana, A.F. (eds.), *Energy, economics, and the environment*. Westview Press, Boulder, CO.
Cromar, K.R., and 19 co-authors. 2022. Global health impacts for economic models of climate change: a systematic review and meta-analysis. *AnnalsATS* 19:1203–1212.

Daly, H.E., Farley, J. 2011. *Ecological economics: principles and applications.* Island Press.

Deacon, R.T., Brookshire, D.S., Fisher, A.C., Kneese, A.V., Kolstad, C.D., Scrogin, D., Smith, V.K., Ward, M., Wilen, J. 1998. Research trends and opportunities in environmental and natural resource economics. *Environmental and Resource Economics* 11:383–397.

De Vogli, R. 2013. *Progress or collapse: the crises of market greed.* Routledge, London.

Dyllick, T., Hockerts, K. 2002. Beyond the business case for corporate sustainability. *Business Strategy and the Environment* 11:130–141.

Ehrlich, P., Ehrlich, A. 1981. Extinction: the causes and consequences of the disappearance of species. Random House, New York.

Eichner, T., Tschirhart, J. 2007. Efficient ecosystem services and naturalness in an ecological/economic model. *Environmental Resource Economics* 37:733–755.

Gao, J., Bansal, P. 2013. Instrumental and integrative logics in business sustainability. *Journal of Business Ethics* 112:241–255.

Hahn, T., Pinske, J., Preuss, L. 2015. Tensions in corporate sustainability: towards an integrative framework. *Journal of Business Ethics* 127:297–316.

Halpern, D., Sanders, M. 2016. Nudging by government: progress, impact and lessons learnt. *Behavior and Science Policy* 2:54–65.

Hausman, D.M. 2008. *The philosophy of economics. An anthology.* 3rd edition. Cambridge University Press., Cambridge, UK.

Hernandez, R.R., and eight co-authors. 2014. Environmental impacts of utility-scale solar energy. *Renewable and Sustainable Energy Reviews* 29:766–779.

Jansson, A.M. (eEd.),. 1984. *Integration of economy and ecology: an outlook for the eighties.* University of Stockholm Press, Stockholm.

Kemeny, P.C., Torres, M.A., Fischer, W.W., Blattler, C.L. 2024. Balance and imbalance in biogeochemical cycles reflect the operation of closed, exchange, and open sets. *Proceedings of the National Academy of Science* 121.

Martinez-Alier,J., 1987. *Ecological economics: energy, energy, environment and society.* Blackwell, Oxford.

Mbak, E. 2010. *Cidanau watershed PES scheme*, Indonesia. TEEBcase, The Economics of Ecosystems & Biodiversity, Indonesia. http://www.teebweb.org

Ripple, W.J., Wolf, C., Gregg, J.W., Rockstrom, J., Newsome, T.M., Law, B.E., Marques, L., Lenton, T.M., Xu, C., Huq, S., Simons, L., King, D.A. 2023. The 2023 state of the climate report: entering uncharted territory. *BioScience* 73:841–850.

Ripple, W.J., Wolf, C., van Vuuren, D.P., Gregg, J.W., Lenzen, M. 2024. An environmental and socially just climate mitigation pathway for a planet in peril. *Environmental Research Letters* 19:021001. DOI: https://doi.org/10.1088/1748-9326/ad059e

Schumpeter, J.A. 1951. *Theory of economic development.* Harvard University Press, Cambridge, MA.

Sterner, T. 2003. *Policy instruments for environmental and natural resource management.* Resources for the Future, Washington, DC.

Turner, G.M., Elliston, B., Diesendorf, M. 2013. Impacts on the biophysical economy and environment of a transition to 100% renewable electricity in Australia. *Energy Policy* 54:288–299.

UN (United Nations). 2009. *System of Nationals accounts 2008.* United Nations ST/ESA/STAT/SER.F/2/Rev.5., New York. 662 p.

WCED (World Commission on Environment and Development). 1987. *Our common future.* Oxford University Press, Oxford.

Index

Printed in the United States
by Baker & Taylor Publisher Services